TO THE
ISLAND
OF
TIDES

Alistair Moffat was born in Kelso, Scotland in 1950. He is an award-winning writer, historian and Director of Programmes at Scottish Television, former Director of the Edinburgh Festival Fringe, and former Rector of the University of St Andrews. He is the founder of Borders Book Festival and Co-Chairman of The Great Tapestry of Scotland. He is the author of *The Hidden Ways: Scotland's Forgotten Roads*.

Also by Alistair Moffat

The Sea Kingdoms: The History of Celtic Britain and Ireland
The Borders: A History of the Borders from Earliest Times
Before Scotland: The Story of Scotland Before History
Tyneside: A History of Newcastle
and Gateshead from Earliest Times
The Reivers: The Story of the Border Reivers
The Wall: Rome's Greatest Frontier
Tuscany: A History
The Highland Clans
The Faded Map: The Lost Kingdoms of Scotland
The Scots: A Genetic Journey
Britain's Last Frontier: A Journey Along the Highland Line
The British: A Genetic Journey
Hawick: A History from Earliest Times
Bannockburn: The Battle for a Nation
Scotland: A History from Earliest Times
The Hidden Ways: Scotland's Forgotten Roads

TO THE
ISLAND
OF
TIDES

A JOURNEY TO
LINDISFARNE

ALISTAIR
MOFFAT

CANONGATE

This paperback edition published in Great Britain, the USA and Canada in 2021
by Canongate Books

First published in Great Britain, the USA and Canada in 2019
by Canongate Books Ltd, 14 High Street, Edinburgh EH1 1TE

Distributed in the USA by Publishers Group West
and in Canada by Publishers Group Canada

canongate.co.uk

1

British Library Cataloguing-in-Publication Data
A catalogue record for this book is available on
request from the British Library

ISBN 978 1 78689 634 6

Typeset in Dante MT Std by
Palimpsest Book Production Ltd, Falkirk, Stirlingshire

Printed and bound in Great Britain by Clays Ltd, Elcograf S.p.A.

For Richard Buccleuch

Contents

Author's Note   ix
Preface: A Short History of Lindisfarne   xiii

PART ONE: TO THE ISLAND

Map: Journey to Lindisfarne   2
1. The Island of the Evening   3
2. The Hill of Faith   14
3. In the Sacred Land   31
4. Soul-Friends   60
5. In the Arms of Angels   91
6. The Quiet Waters By   139
7. Wandering   160

PART TWO: LINDISFARNE

Map: The Holy Island of Lindisfarne   184
8. On the Island of Tides   185
9. The Winds of Memory   209
10. Duneland   230
11. When God Walked in the Garden   248
12. Crossing the Causeway   276
13. The Rock   283

Epilogue: Godless   299

Acknowledgements   307
Index   309

Author's Note

One of my most treasured possessions is a small cache of letters. Written in a looping, spidery hand with sentences that turn corners up the sides of pages before abruptly dipping overleaf, they are full of criticism and advice. The writer was a retired librarian and borrower of all my books but I never knew his or her name. Received over a four-year period, all of the letters were signed A Reader and no return address appeared at the top of the first of many pages. Not that there was room for one.

The last letter was never posted. It arrived inside another envelope with a compliments slip from a care home in Berwick-upon-Tweed. I took it that my critic had died before the letter could be dispatched and something prevented me from calling the care home to ask for a name. He or she had not wished to give it and I felt I ought to respect that.

Most of the criticism was a helpful mixture of pointing out blunders, a wrong date or mistaken identity and suchlike, and there were occasional comments on inaccurate use of language and poor style. 'Posterity is not an interchangeable term for history' and 'using a dash is simply slovenly' or 'try not to over-use the ablative absolute at the beginning of a paragraph'.

His or her advice was to try to understand better the

importance of place in history and to get out from in front of my screen and visit the sites of important events or where important people passed their lives. One letter surprised me by suggesting (or maybe insisting) that I should read the opening chapter of Daphne du Maurier's *Frenchman's Creek*, 'a delicate and enchanting evocation of place and how it has been seen differently over the centuries'.

Before I began work on this book, I took this unexpected advice and re-read *Frenchman's Creek*. I was indeed enchanted once more. At the peak of her powers, du Maurier wrote about the Helford River mouth on the Cornish coast and the inlet that gave her novel its title. Almost cinematic in its imagery, the opening chapter is intensely atmospheric, a world of winds, tides and a silence broken only by the call of nightjars. Du Maurier moves seamlessly between Restoration England and the twentieth century, establishing Navron House, the central location, and then, six pages in, the story begins with the reader transported almost trance-like back to the past. It is magical and masterly.

What my critical correspondent from Berwick was suggesting, if that is not too mild a verb, was that a deeper understanding of place would enhance the dry recital of dates and events that form the framework of a historical narrative. And more, it would make a powerful link between the present and even the deep past. Fernand Braudel, the great French historian, wrote of the *longue durée*, the notion that peoples in similar geographical, social and economic circumstances maintain habits of life and mind over vast reaches of time. If a sense of a place, its atmosphere as much as its topography and climate, could be caught and understood, then history would come alive.

About twelve years ago I was making a TV series on the history of Tyneside, a part of the world I am very fond of, and we were filming at Durham Cathedral. Setting up to shoot a sequence around the shrine of St Cuthbert, we were joined by the stern lady who ran the administration of the great church. I imagine she was checking that we were behaving ourselves. When she pronounced Cuthbert the greatest English saint, I felt compelled to point out that he had been born and raised in what is now Scotland. Before I could add that the border did not exist in the seventh century, she exclaimed, 'Nonsense! Utter nonsense,' and stormed off back down the nave. We packed up our gear, eventually completed the series and I moved on to another project. But I never forgot that exchange. Thirteen centuries after his death, Cuthbert's exemplary life still attracted fierce loyalty, even passion, to say nothing of contested history. One day I would find the time to write about him.

In 2017, that day came when I began to think seriously about this book. I am no Christian, but sainthood and how it was and is achieved interests me very much. Heroes and heroic actions I can understand and admire, but the journey from ordinary life to an elevated, demi-godlike state is something I wanted to know more about. Cuthbert lived from approximately 635 to 687, a time of seismic cultural shifts in Britain, the period often known as the Dark Ages because of the dearth of extant written record. As Anglo-Saxon invaders became settlers who supplanted and submerged Celtic society over much of lowland Britain, languages were exchanged and enduring identities forged. It was a time of great, largely unreported turmoil.

Cuthbert is an Anglian name and he may have been born into the first or second generation of a landed family who settled in the Tweed Valley. I knew that around 651 he entered the monastery at Old Melrose, where he took holy orders. After some years there, he travelled east to the Holy Island of Lindisfarne, became prior of the monastery and eventually bishop before dying in his hermitage on the little island of Inner Farne. I thought that if I could understand something of his journey through life and, as importantly, through the landscape, then I might be able to move close to a sense of the man who walked with God.

I wanted to go to Lindisfarne, a deeply atmospheric place I have come to love, having visited it half a dozen times over the years. Now, its peace and spirituality seem even more to be valued, as political chaos swirls around us, as we blunder through the wilderness of a divisive and bitter world that seems at times to be irretrievably broken. And so I decided on what would become two journeys.

To be closer to Cuthbert, I wanted to walk where he walked and spend time in the places where he passed his life. I would begin at Old Melrose and travel down the banks of the Tweed to its confluence with the River Till before following Cuthbert inland to the moors of northern Northumberland. Once over the Kyloe Hills, I would find the coast and cross the causeway to Lindisfarne to complete what might be called a secular pilgrimage.

I also wanted to make the journey for myself. While walking in the shadow of Cuthbert and trying to understand his faith and the extreme lengths he went to in pursuit of piety and purity of thought and deed, I would also think about my own life. The contrast between his asceticism and beliefs and my lack of either could not have been more stark, but I hoped very much that I might learn something from Cuthbert.

And, dear Reader, I am sad not to be able to point out to you that no paragraph in this introduction begins with an ablative absolute, although I do recall that you also disliked sentences that begin with conjunctions.

Preface

A Short History of Lindisfarne

Having rattled through the featureless flatlands of the Fens and past the remains of the old industrial heartlands of South Yorkshire, travellers on the London train bound for Edinburgh pass a place of great majesty. As the carriages slow and glide through the small station, many look up from their newspapers or phones to gaze at Durham Cathedral. The mass of its towers and great nave perch on a river peninsula above the Wear and they speak of faith, of continuity, of solidity and of half-remembered history.

The interior of the cathedral is as awe-inspiring as the exterior. Immense pillars carry the soaring roof and lead the eye down to the high altar where the glorious stained glass of the rose window glows above it. Behind all of that splendour is what made it possible. On the floor is a slab of rust-coloured marble with the name 'Cuthbertus' inscribed on it. Durham Cathedral was raised on Cuthbert's bones. Such was the power of his cult and the love believers felt for this gentle, vulnerable, deeply devout man that the towers of the great church soared to the heavens to celebrate his life, his miracles and his ability to inspire a simple and profound faith.

Beyond Newcastle, the train edges ever closer to the North Sea coast, as the lovely village of Alnmouth comes into view.

Not far south of Berwick-upon-Tweed, a low, sandy island can be seen, its farthest point punctuated by a steep rock topped by a fairytale castle. Nearby a small cluster of roof-tops huddle around the ruins of a church.

Lindisfarne's beauty is quieter, less dramatic than Durham's but its innate spiritual power is immeasurably greater. Saints walked on the island, miracles were seen, and between 685 and 687 Cuthbert was its bishop, master of a vast patrimony of land and glittering wealth. The sandy island was the centre of early Christianity in seventh-century England, a place where great faith was forged in the fires of privation, prayer and personal sacrifice.

Place-names can be emblematic, compressed nuggets of forgotten history, and the mouth-filling sensuousness of Lindisfarne may have a tale to tell. It seems that the Romans knew it as Insula Medicata, and that may have referred to an island where plants grew that were useful for making balms, poultices or decoctions. Dialects of Old Welsh were spoken down the length of Britain before and after the Roman province of Britannia, and the native name of Ynys Medcaut derived directly from the Latin name.

Throughout the fifth and sixth centuries Angles and Saxons sailed the North Sea from Scandinavia and what is now Germany and Holland to settle on the eastern shores of Britain. The kings of Lindsey ruled over the low-lying lands south of the Humber and their people knew themselves as the Lindisfaras. A band of them appears to have colonised Ynys Medcaut, perhaps building a defensive stockade on the rock where the castle now stands.

When Gaelic-speaking monks came to the island in the seventh century, they may have brought the second element of the island's name. *Fearann* means 'a piece of land', and Lindis-fearann was coined: the land of the Lindisfaras.

Led by Aidan, twelve monks came from Columba's famous monastery on Iona at the invitation of the Northumbrian King Oswald. As a young man he had been exiled and sought sanctuary on Iona, where he was baptised. Having won a decisive victory over a coalition of native British kings at Heavenfield, near Hadrian's Wall, Oswald established his capital place on the sea-mark rock at Bamburgh. And so that support and resources were readily available to the new community, Aidan decided to build the first church and its monastic cells close by, on Lindisfarne.

With King Oswald translating his Gaelic for the pagan Anglian settlers, Aidan began the process of conversion. At Old Melrose on the River Tweed, then part of Northumbria, he founded another monastery. In 651 Cuthbert entered holy orders there and his long journey to Lindisfarne and saint-hood form the core of the narrative that follows.

The Church in Britain was riven with controversy and in 663 Oswald's brother, Oswy, convened a synod at Whitby to settle all arguments. Roman practices were preferred over the customs of the Celtic Church and Bishop Colman of Lindisfarne felt compelled to return to Iona. The story of these difficult times is concisely told (albeit not entirely objectively) by the greatest scholar of early medieval Britain, Bede of Jarrow, and he wrote of Cuthbert's exasperation with those monks on the island who would not conform to Roman rules.

Lindisfarne became a place of immense richness. To provide an additional and prestigious focus for the cult of Cuthbert, Bishop Eadfrith painted one of the greatest works of art in our history, the *Lindisfarne Gospels*. Around the year 700, this glorious object was made on the island with extra-ordinary and unexpected skill. Not only would there have been a well-furnished library and plants grown to make pigments, but there were also monks who were able to create a tooled and bejewelled metalwork cover and who could bind the pages of calfskin. The great gospel book is an unlikely achievement for a community who lived in leaky wooden huts on a windy, sandy little island off the cold shoulder of the Northumbrian coast.

This beautiful book and other gilded treasures attracted unwelcome attention. In 793 the Vikings attacked the monastery, one of the first raids on the British mainland. Known as the Sons of Death, these pagan warriors eventually forced the community to abandon Lindisfarne. For about one hundred and fifty years, the Congregation of St Cuthbert wandered northern England, carrying with them the coffin, relics and treasures of St Cuthbert. In 995 all three were finally enshrined at Durham and by the early eleventh century the great cathedral had begun to rise.

In 1083 the old Anglian congregation was replaced by Benedictine monks and all links with Lindisfarne were in danger of being severed. The powerful prince-bishops of Durham determined to re-forge the relationship with the island where Cuthbert had spent the last years of his life. A priory was founded on the site of Aidan's original monastery and its ruins dominate the modern village. The new church was built opposite the parish church of St Mary. The priory had been raised over the ruins of an earlier Anglian church. Very little can now be seen of the monastery where Aidan, Cuthbert and Eadfrith walked through their exemplary lives.

When Henry VIII's dynastic difficulties persuaded him to dissolve England's monasteries, Lindisfarne was quickly deserted, despite its venerable origins. After 1537, the island became a naval base in the sporadic wars with Scotland and stone was robbed out of the priory to build a fortress on the castle rock. Nevertheless, much of the priory church remained intact as late as the 1780s, when it was visited by antiquarians and artists. By the 1820s, the central tower and the south aisle had collapsed and much stone was carted off to build houses in the village and elsewhere.

In the last two hundred years, Lindisfarne's fame and spiritual magnetism have been reborn. Now thousands of visitors cross the causeway to visit the holy island, its priory and castle. As many of these modern pilgrims walk in the footsteps of great saints, they see more than ruins, feel more than a sense of the long past. They come because the island is still a place of spirits, a place where they can hear the whispered prayers of Cuthbert and the echo of psalms that were once sung under these huge skies.

PART ONE

TO THE ISLAND

Journey to Lindisfarne

Berrig

Old Melrose

Brother' Stones

Bemersyde

Dryburgh Abbey

Roxburgh Castle

Kelso

Whitmuirhaugh

Mouth of the Till

Etal

Ford

Causeway

Cuddy's Cave

Holburn

St Cuthbert's Cave

The Island of the Evening

Most of a lifetime ago, in the summer of 1965, I walked across the causeway to Lindisfarne. Only just turned fifteen, I was one of three schoolboys intent on adventure, or at least a change of scene, something different from the dull school day routine that had just come to an end. Six glorious weeks of summer holiday were opening before us and we had decided to celebrate our freedom by travelling south-east from Kelso, our home town, to cross the border to the Northumberland coast and enjoy all the exotic differences of England.

I have no memory of the journey except for two moments, one misty, the other sharp. We must have walked on the roads, much less busy in those days, my companions and I, because my vague recollection is of pitching a tent, borrowed from the Boy Scouts, on a riverbank near the little town of Wooler. Midgies are what sticks in the mind, and everywhere else.

The sharper memory is of crossing the main line between Edinburgh Waverley and London King's Cross at Beal. The white level-crossing gates were open and, it being a hot day, we stopped at the red brick station house on the far side to ask for a glass of water. This was a time long before anyone thought of bottling the stuff. In what I imagine must have been an interval between trains, the signalman/gatekeeper was in his garden tending to conical trusses of canes that were covered in the pale colours of sweet pea blossom. It was a still day and the air was thick with their scent.

When he came out of his back kitchen with a jug of water and glasses, this genial man had also put a large jar on the tray. Saying something like, 'You lads won't have tasted this before,' he unscrewed the top and pulled out what looked like a twig with small blobs of green stuff stuck onto it. 'This is samphire, edible seaweed,' he explained, and drew the twig through his teeth, then pointed it towards the seashore. 'Down by the causeway there's plenty, but not many know what to do with it.'

The three of us probably exchanged glances, and I went first. It tasted of salty vinegar. A mild grimace must have escaped because the signalman laughed as he took the big jar back into the kitchen. 'If you are wanting to cross' – he waved his arm in the direction of the low sandy island we had seen in the east – 'then you'll need to get a move on.' Of course, it had not occurred to any of us to check the tide times, probably because none of us knew how.

A few minutes later, having climbed the low hill where the hamlet of Beal clusters, a wide vista opened. Below us was the arrow-straight tarmac causeway, maybe a bit more than a mile long, and beyond that, the Holy Island of Lindisfarne, its village huddled around the ruins of the priory and the miniature castle rock further in the distance. To the north lay a long run of undulating sand dunes, and beyond all of that the shimmer of the North Sea and its endless horizon.

We hurried down the slope and began marching across the little black road that seemed built on sand, its edges fraying. On either side, the long, wet sands seemed to merge seamlessly with the sea, and there was no telling if the tide was coming in or going out. We were carrying hefty packs and moved as fast as we could towards the white refuge box that stood on stilts about halfway across. Its purpose was to rescue idiots like us, but when we reached it, we still could not tell if the tide was coming in and, being daft boys, we pressed on regardless. There was no other traffic on the causeway, wheeled or on foot, but of course the significance of that occurred to none of us.

By the time we were halfway between the refuge box and what seemed to be dry land (or was it?), there was no doubt. On both sides we could see the tide rising, what looked like a low wave running across the sands towards us. We began to jog, and when the sea washed across the tarmac we started to run. For the last hundred or so yards, we were forced to slow to a wet walk as we first splashed and then waded. And then the road rose up almost imperceptibly, suddenly dry, and we stopped, panting, shuffling off our rucksacks, then turning to look behind us as the tide raced across the black tarmac and rose up the stilts of the refuge box.

Landlocked boys raised forty miles inland amongst the meadows and cornfields of the Tweed Valley, we had no idea of the elemental power of the sea. No matter how far we strayed from home in Kelso on a summer evening, we could always walk back, always get home, even if there was a telling-off waiting. But here we were sea-locked; the tide had cut us off from the world, and nearly done worse. In moments, Lindisfarne had become an island, and we were stuck on it. Not so much bad planning as no planning.

Our boots and socks were soaking and there was nothing we could do about that except squelch along the road. I remember in those days I wore a green hooded anorak with a large, horizontally zipped pocket along the front. It was the sort you had to pull over your head. Under it my shirt was soaked with sweat, but the road to the village was long and sweeping. Perhaps I would dry off before we reached it and could ask when the tide was due to go out. By the time we started our damp trudge to the village, it was late afternoon.

At the age of fifteen I was as tall as I am now (or was until I started recently to shrink): six foot and a bit. So I decided that since no one knew us on this sudden island, we should go into a pub and order drinks, as well as ask about the tide times. My grannie, Bina, used to enjoy a glass of a sweet stout called Mackeson, and when my mum wasn't looking she gave me a taste of it – and for it. It must have been a mixture of my Scottish accent and some nervousness, but when I ordered it the landlord put a large bottle of something called Double Maxim on the bar. Brewed by Usher Vaux, it was a brown ale that turned out to be revolting, bitter and sickly at the same time. Because we could not risk being found out as underage drinkers (we had seen a policemen in the village), we had to pretend to like it and between the three of us we managed quickly to swallow this vile concoction. Worse was to follow when the barman told us that the causeway would not open again until about 10 p.m.

By the time we left the pub, feeling a little tipsy, having flung down what turned out to be a strong ale (like Newcastle Brown Ale), our spirits plummeted when we saw that the village shop, the only possible source of food, had closed. I went back into the pub and bought three packets of crisps. Sitting on a bench near the ruins of the priory, having untwisted the blue packets of salt, shaken it over the crisps and then shaken the bags, we devoured our meagre supper and agreed that we should probably not cross the causeway in the half-dark and then blunder about trying to find some-where to camp on the mainland. Better to stay on the island.

But where? The dunes to the north looked promising. When we walked out of the village, past some bungalows that seemed the epitome of comfort, their larders doubtless bulging with food, we saw a sign by the car park that sent our hungry spirits sinking ever lower: 'By Order of the Cheswick Estates, No Camping'.

Why we decided to walk to Lindisfarne, or whose idea it was, is mercifully lost to memory. And looking back now, having helped raise children of my own, I am amazed that our parents allowed us to go so far from home without any adult accompaniment, or even any means of contacting them. My parents did not have the phone and I doubt if my companions' families did either. It never occurred to me to ask them. In many ways the 1960s were more innocent times, but those degrees of trust and freedom still surprise me.

I recall that the year before, the three of us had been inspired by President Kennedy's challenge to Americans to improve their fitness. His measure was simple. If you could walk fifty miles in twenty-four hours, you could claim to be fit – and in the summer of 1964, the year after Kennedy died in Dallas, we did it. Setting off in the middle of the after-noon, we walked through the night to Berwick-upon-Tweed from Kelso and back again with only a brief rest in a bus shelter. Somewhere near Norham, where the road bends around the entrance to a farm, we sat down on a grassy bank to eat the last of our sandwiches. I must have fallen asleep, for when I woke in the darkness, my friends had disappeared. Leaving me alone. As I took a few, fearful steps down the long and lonely road home, they jumped out from behind a tree. Perhaps because we managed that without mishap, our parents allowed us to do another long distance walk – although if we had told them of our antics on the causeway, it would have been the last one.

While the reasons for choosing Lindisfarne have long fled into the darkness of the past, there were plenty of grumbles about being on a bloody island with no food and no choice but to break the law by camping. As we approached the dunes across a broad, open area, a light came on in my head. If we did not camp, but only slept out, then the policeman we had seen could not arrest us. We were not camping, only sleeping. Nothing about that on the sign. And anyway, it was a mild, even balmy, night, with no breeze to speak of. Suddenly it seemed that our troubles were turning into an adventure.

By the time we passed the last of the fenced grass parks and their small, snuffling herd of cows, the sun had set away to the north-west, slipping behind the Cheviots, and gloaming fell. Beyond the farm fields lay the line of sand dunes and as the bright day gave way to full moonlight, we were able to find a path through the undulations and the clumps of spiky marram grass. Looking at a map, more than fifty years later, I think we must have walked towards the north-east corner of the island. So that we could keep lookout – for the police, and perhaps the enraged owner of the Cheswick Estates, or whoever had put up the sign – we searched the horizon for the highest dune.

For our fevered boyhood sense of conspiracy had persuaded us that we would definitely be hunted, tracked down. In the village, all eyes had been on us. From behind curtains, behind the bar, on street corners, as we splashed across the causeway, we had been noticed. Definitely. Strangers were in town. And even now a group of vigilantes (we were well versed in the terms of TV westerns) were probably being deputised by the village policeman to form search parties to find and arrest these intruders. In these parts, Scots had long been thought of as suspect. Even if our lawyers could argue down the charge of illegal camping on a technicality, we could still be convicted of under-age drinking, a serious offence, especially in England, probably. At any moment sirens might wail, blinding searchlights clang on, and we would be hauled off to spend a night in the cells. In the dread phrase, my parents and Bina would be black affronted and my sisters would never let me forget it.

When we scrambled up the highest dune we could find, each step pushing down small avalanches of sand, we saw that there was a shallow plateau behind it, encircled by a ridge of sand held together by marram grass. Hunkering down there, we would be out of sight from below. One of us had found some Bassett's Treacle Toffees in an ignored pocket and we sucked slowly on the sugar, trying to make each one last.

By some forgotten process, probably because it was all my fault, the under-age drinking in particular, it was decided that I should take the first watch while the other two laid out the ground sheet of the tent (was that camping?) and wriggled into their sleeping bags.

Using the rim of the sand dune like an imaginary rampart, I lay on my front looking south, the direction trouble would come from, and I scanned the twinkling lights of the sleeping village and the dark silhouette of the castle, presumably the stronghold of the fearsome owner of the Cheswick Estates and his henchmen. Far beyond, it seemed, I saw the sweeping beam of a lighthouse playing across the sea. At the time I did not know it was the Longstone light on the Farne Islands, the place from where Grace Darling and her father set out in their famous recue of shipwrecked sailors. Idiotically, I wondered if the light showed up our position to the bands of desperadoes who were roaming in the gloaming, hunting for us.

After a time, I turned to see that my companions were lying still, fast asleep. It was a balmy, windless night, well lit by a full moon, the sort of half-dark sometimes called the summer dim in the north of Scotland, and as thoughts of marauding bands of policemen gradually faded and a silence settled, I found myself looking, almost hypnotically, over the endless wastes of the North Sea and listening to the wash of the waves. To the south, there was enough light in the sky to see the outline of Bamburgh Castle, although I was not sure what it was. Little by little, the dune became less like a rampart and more like a high vantage point, a place from where I could see the sky, the sea and the land. My watchnight cannot have lasted more than an hour or perhaps two, but I remember it vividly.

Perhaps I am reading history backwards, but I think on that moonlit, silent night a peace descended on me, something I had not felt before. It may be that after the excitements, exertions and daftnesses of the day I simply relaxed, realising that no one was looking for us, that we were alone out in the dunes. But if that was so, then surely I would have fallen asleep like my companions. Instead, without knowing it, I think I was keeping vigil, unconsciously allowing the spirits of that place to swirl around me and release a sense of inner calm as I looked out over Creation. This may sound highly unlikely for a fifteen-year-old boy on the threshold of life, but looking back across the years I believe it to be true. It turned out to be a beginning of sorts.

Ignorant of the phrase, I sensed then that the island had a powerful *genius loci*, was a place of spirits. And each of the many times I have returned in the last fifty years, knowing more, I began slowly to realise that for me Lindisfarne might be more than beautiful, atmospheric. It might be the saving of me.

*   *   *

Now nearing the end of my seventh decade, I need to face some actuarial facts. I may have ten more summers of active, relatively healthy life in front of me, if I am lucky. But to enjoy them – and face the end when it comes – I need to change what is in my heart and soul. With much more past than future, I have found myself dwelling not so much on good memories as on the darknesses, all that went wrong, bad mistakes I made, people I hurt, people who wronged me or ignored me. In the half-light of early morning, drifting in and out of sleep, the ghosts of disappointments, mistakes, slights and regrets flit through my mind too often and I fear I am becoming more and more bitter rather than wise, accepting and forgiving, clinging to the hurts of the past rather than savouring the everyday joys of the present. What I need to find is some peace of mind and it occurred to me that if I took myself to Lindisfarne for a time, I might find some in that place of spirits. After all, the old name Ynys Medcaut, from Insula Medicata, means 'the isle of healing'.

Buried for more than a century on the island, and once its bishop, Cuthbert was the great saint of the north. Durham Cathedral and the vice-regal powers of the medieval prince-bishops are his direct legacy and his tomb is still much venerated. But I believe that his spirit has never left Lindisfarne. As Cuthbert and his ascetic companions watched the sun rise over the grey chill of the North Sea and listened to the bleak music of the winds and the waves, they knew that all of the elements of the world were writ large before them. Somewhere in the huge skies above the island, God was moving and they prayed that somehow His spirit would descend and be revealed in this place that stood apart from the world.

Probably born and raised as the younger son of Anglian lords in the hills of Lauderdale in the Scottish Borders, Cuthbert came to the monastery at Old Melrose some time around 650 or 651. While tending sheep one starry night, the boy had looked out to the east and seen a vision, believed to be the ascent of the soul of St Aidan from his church on Lindisfarne. Much moved by this, Cuthbert rode to meet the monks in the loop of the River Tweed at Old Melrose and asked to take holy orders. When Prior Boisil died a few years afterwards, the shepherd boy succeeded him. And then, perhaps in 676, he laid down the cares of office and began life as a wandering hermit, seeking to move closer to God. Two caves near the Northumberland coast are associated with his exemplary life.

But by Christmas 686, Cuthbert knew that he was beginning to die and, having been reluctantly appointed Bishop of Lindisfarne two years before, he put down his cope, pectoral cross and staff to return to the purity and peace of his solitary life in the cold hermitage on Inner Farne. A short time later, he died.

As I have said, I am no Christian, and nor does mortification of the flesh hold many attractions, but I would like to seek something of the peace Cuthbert sought at the end of his life. My DNA ancestry is Northumbrian, as his must have been, and my people sailed the North Sea in the fifth and sixth centuries from Denmark and southern Sweden to settle on the eastern coasts and in the river valleys of Britain just as his people did. I was born and raised only a short distance from the hills where he tended sheep. I know that thirteen centuries separate us, but perhaps I could walk the holy ground of Lindisfarne in Cuthbert's footsteps; perhaps some of the eternal spirits of that magical place would speak to me.

I began to plan a secular pilgrimage, to walk from the ancient monastery at Old Melrose to what I had come to think of as the Island of Tides, and on my way I would visit places where Cuthbert prayed, preached and wrought miracles. As well as trying to understand something of the harsh lives of these leathery old saints and their efforts to know the mind of God, I knew during this journey of endings and beginnings I would also come across some glorious, life-enhancing colour – from the disciplined joy of the *Lindisfarne Gospels*, the glittering culture of the kingdom of Northumbria, the scholarship of Bede of Jarrow, as well as the story of Cuthbert, one of the first and perhaps the greatest saint of Britain. And maybe I would discover on the Island of Tides that the tides in the lives of men are not so very different.

2

The Hill of Faith

It was a day of high summer. A cloudless July sky, perfectly still, without even a breath of a breeze riffling the topmost leaves of the old sycamores on the ridge above the track. In the Tile Field, about half a mile below our farmhouse, all the sheep and cattle were lying down and the only animals I could see moving were the four horses who graze the field beyond them. They are at a DIY livery at the neighbouring farm and each morning one or two women come with bowls of hard feed and a barrow to pick up their muck. It was so quiet I could hear them talking. It is astonishing how sound carries across our little valley, and this morning I could clearly hear the bleat of sheep in the fields above Brownmoor, as well as the distant purr of a quad bike as the shepherd went up onto the southern ridge to check on the flock about a mile to the south.

As Maidie, my West Highland terrier puppy, and I walked along the track by the old wood, the women must have been three-quarters of a mile away. Not far enough for Maidie. A very territorial terrier, she had stopped when the horse-women began to speak, pricked her ears and barked loudly at them. How dare they! Who are they? This set off the collies up at Brownmoor, and then moments later, across the stillness of the morning, I heard the big, booming bark of Chiquita at Burn Cottage down at the bottom of the Long Track that leads to our farm. She is a huge Newfoundland. Doggy stereo rang round the hills. But it soon calmed down and we walked on through the waking landscape, stopping and staring at all our familiar glories.

We live on a small farm in the Scottish Borders. At least I think we do. I have never been sure what to call the little bit of the planet we own and look after. At eighty acres, it is bigger than a smallholding or a croft; I would wince if anyone called it an estate, but it seems too small to be a farm. We do not grow any crops except grass, and my wife manages about twenty-five acres of pasture to breed horses. So, I guess 'small farm' is the least inaccurate description. Below the house we have gradually built up a stable block that now has ten loose boxes, a hay barn and a tack room, and by the track that leads down to the lane and from there to the main road we have added more outbuildings, an arena and barns. One of these is my office, a large room overflowing with books and runs of periodicals that have helped me write mainly Scottish history over the last twenty-five years.

Perhaps one of my most eccentric possessions is a series of annual volumes published by the Berwickshire Naturalists' Club. Founded in 1831 by Dr George Johnston of Berwick-upon-Tweed, it is the oldest active natural history field club in Britain and its object (what would now be called a mission statement) is pleasingly limited. Members are interested in 'investigating the natural history of Berwickshire and its vicinage', while the club badge carries the figure of a wood sorrel, Johnston's favourite flower, and the motto is *Mare et Tellus et quod Omnia, Coelum* – 'The Sea and the Land and what covers all, the Heavens'. Wonderful. Especially the use of the obscure 'vicinage' instead of vicinity.

From the BNC, I discovered that in the summer of 1930

a group of members had visited what they reckoned to be the ancient village of Wrangham between Kelso and St Boswells. Their guide, Rev. W.L. Sime, pointed to a row of ancient ash trees at Brotherstone Farm, saying that it grew where the village once stood. Andrew Armstrong's map of the Borders made in 1771 also plotted its remains at the farm. This immediately caught my interest because I had been reading the Anonymous *Life of St Cuthbert* in the excellent translation by Bertram Colgrave, a scholar at the University of Durham. There are three eighth-century versions of lives of Cuthbert, two written by that more venerable and very great scholar Bede of Jarrow, but the earliest is the anonymous text almost certainly written by a monk on Lindisfarne between 699 and 705, only twelve or at most eighteen years after Cuthbert died.

The first *Life* is also more richly detailed than the others and contains what sounds like testimony from monks who knew Cuthbert. The account of his miracles and deeds feels at once more personal, more authentic, less formulaic, less political. It seems that Cuthbert was of Anglian rather than native Celtic descent (his name itself suggests that); the son of a landed, perhaps noble family, for the young boy was sent to be fostered, a common practice in early medieval Britain and Ireland and a means of extending bonds of loyalty and obligation amongst ruling elites. The prime purpose of both Bede's and the Anonymous *Life* was to establish Cuthbert's cult of sainthood and recount the miracles he wrought with God's help. Here is a passage from the earlier *Life* that also imparts a little biographical detail:

At the same time the holy man of God was invited by a certain woman called Kenswith, who is still alive, a nun and a widow who had brought him up from his eighth year until manhood, when he entered the service of God. For this reason he called her mother and often visited her. He came on a certain day to the village in which she lived, called Hruringaham; on that occasion a house was seen to be on fire on the eastern edge of the village and from the same direction a very strong wind was blowing, causing a conflagration.

Cuthbert fell to the ground and began to pray, and miraculously the wind began to blow from the west and the rest of the village was saved. Over time, the place-name of Hruringaham was rubbed smooth into Wrangham and, to add to the findings of the Berwickshire Naturalists, archaeologists have more recently detected the remains of an ancient village near the modern farm of Brotherstone on the slopes of the Brotherstone Hills above the road between Kelso and Melrose.

Here is another description of a miraculous event from the Anonymous *Life*:

On another occasion, also in his youth, while he was still leading a secular life, and was feeding the flocks of his master on the hills near the river which is called the Leader, in the company of other shepherds, he was spending the night in vigils according to his custom, offering abundant prayers with pure faith and a faithful heart, when he saw a vision which the Lord revealed to him. For through the opened heaven – not by a parting asunder of the natural elements but by the sight of his spiritual eyes – like blessed Jacob the patriarch in Luz which was called Bethel, he had seen angels ascending and descending and in their hands was borne to heaven a holy soul, as if in a globe of fire. Then immediately awakening the shepherds, he described the wonderful vision just as he had seen it, prophesying further to them that it was the soul of a most holy bishop or of some other great person. And so events proved; for a few days afterwards, they heard that the death of our holy bishop Aidan, at that same hour of the night as he had seen the vision, had been announced far and wide.

What Cuthbert saw, and whether or not it really was a miracle rather than a meteorological event, seemed much less important to me than where he saw it. From all that I had read and researched, it was clear that he had been raised at Brotherstone/Wrangham, not far from the junction of the River Leader with the Tweed, and that he had had a remarkable, transcendent experience, probably in the Brotherstone Hills.

In 2000, I first became interested in Cuthbert when I was writing what I hoped would be a definitive history of the Scottish Borders (it ignored the border for the early part of the story) and I knew that up on one of the Brotherstone Hills there were two impressive standing stones known, of course, as the Brothers' Stones. A third had been raised some way down the eastern slope and called the Cow Stone. I wondered if that place of even older, prehistoric sanctity was where Cuthbert had been tending his flocks and where he had seen the angels and Aidan's soul ascend.

I had never been up to the Brothers' Stones and so, having fed, watered and walked the dogs that sunny morning in July, I pulled on my boots and drove over to Brotherstone. Or at least I thought I did.

Having crossed the magnificent new bridge over the Tweed and then the much older and narrower bridge over the Leader, I drove east towards Kelso before turning left up a short farm track. There was no sign, but the Ordnance Survey suggested this was Brotherstone and glowering above the steading I could see the south-facing cliff of a steep crag. Having rung the doorbell of the farmhouse and had no reply except the furious barking of a very angry guard-dog, mercifully behind the back door, I walked over to a courtyard of cottages. A lady assured me it was OK to park and she would let the farmer know I was planning to walk up to the stones.

Every fence seemed to be electrified and so I carefully ducked under the wire at all of the gates. It was a thick gauge intended to give straying cattle a real jolt. Below the crag I found what looked like a warren of fox holes, or maybe a badger sett. Whatever creatures had excavated the rich, red earth, they had not troubled to seek the cover of gorse thickets or even long grass, and by the dyke lay the eviscerated carcase of a lamb, recognisable only by the largely untouched head, its clouded eyes bulging. Odd.

I skirted the crag and climbed up to a wide ridge of rough grazing made into large parks by long runs of drystane dyking. In front of each was a low electric wire and when I looked for a gate leading me in the direction I wanted to go, there seemed to be none. Perhaps this was a farm boundary. By this time the sun had strengthened and I was regretting not wearing a hat. When I came upon a tumbled section of dyke, I crawled under one electric wire and scrambled up over the stones to see another on the far side. The ground beyond it was obscured by tall nettles and so I had no means of judging how much of a drop it was. But did I really want to climb back down and carry on looking for a gate? When I jumped down, my right foot glanced off a hidden stone and I was lucky not to twist my ankle or worse. This gentle walk up a low hill was turning into a business.

When I reached what I reckoned to be the summit, there were no standing stones to be seen anywhere, only another slightly higher summit about two hundred yards further east. When I reached it, more disappointment waited, more head-scratching, more bewildered consultation of my map. I tried to locate a strip of sitka spruce below me and relate it to the route I had taken. My map was from the old Pathfinder series, about thirty years old and sitkas grew quickly. Was it too old to show the strip? And then it dawned. The unnamed farm where I parked could not have been Brotherstone. Instead of telling her I was going up to see the stones (she must have thought I was taking the scenic route), I should have checked with the lady at the cottages that I was in the right place.

As I marched back downhill and slid and scrambled under more electric fencing, very warm by now, with flies buzzing around me, trying not to become bad-tempered, I remembered another spectacularly bad piece of map-reading more than fifty years before, one that made me smile.

In the mid-1960s orienteering was a new sport in Scotland, imported from the forests of Scandinavia, and at Kelso High School we had Mr Climie, a real enthusiast. Having taught us the basics of map-reading with a Swedish Silva compass, we were then told to set off in search of a series of controls or check-points hidden in the Bowmont Forest, near Kelso. It seemed like good fun, very different, thinking and running at the same time. For the 1965 Scottish Schools Championships we went up to the vast forests around Aberfoyle in Highland Perthshire. The idea was to run around a course in the correct sequence and have your map time-stamped by the officials at each control. The fastest team would be the winners.

In our team of four I was last to go and almost immediately made a catastrophic error. I read my compass bearing

180 degrees wrong and ran round the course backwards, arriving at the finish, which turned out to be the start. My three team-mates had made good times, but to win we needed all four competitors to complete, even though my time would be terrible. They shouted to me that there was only forty minutes to go before the course and the competition would shut down. I turned round and set off again, knowing exactly where all the controls were, having already visited them – in reverse order. I re-appeared at the gate into the field where the finish was with only a few minutes remaining and, responding to the agitated cries of my team-mates, I sprinted home and we won – just.

Having discovered the sign for the real Brotherstone Farm, I finally managed to park in the right place and began once more to walk uphill. A very rough and pitted track ran off to the west, and so I ignored it and climbed a gate to join another that ran to the north-east, the direction I wanted to go. Or so I thought.

Almost immediately I was met by a wall of very dense and prickly gorse bushes. Having turned to the west, I then ran into another insurmountable obstacle: the sheer face of an old sandstone quarry. It seemed that Cuthbert and the Brothers' Stones were working hard to keep their secrets. In fact all of my journey to Lindisfarne would turn out to involve mistakes and more than a little effort, emphatically not a tour guided by a handbook or a reliable map, either paper or cerebral. I had guessed that 'pilgrimage' would probably involve adversity, but not that most of it would be self-inflicted.

When I began to wade through waist-high willowherb and nettles to the side of the quarry, the sun had climbed high in the sky and a cloud of flies was circling around my damp hair and forehead. But then the weeds and the gorse thinned and the walking suddenly improved as I breasted the ridge above the quarry. And there in the distance, on the crown of the easternmost of the Brotherstone Hills, clearly silhouetted against a cobalt-blue sky, was a tall standing stone.

A well-used track, rutted by wheeled vehicles, led up to the stone and gradually I saw that it was the same one that had seemed to swing away out to the west. It had turned towards the sitka spruce plantation and then aimed directly at the summit of the hill. The walking was easy and, as I climbed, the vistas on all sides began to reveal themselves. They were panoramic, long views to all points of the compass, some of them more than twenty miles. A breeze began to cool me, the flies fled and I walked the last few hundred yards in a good rhythm.

As I neared the summit of the hill, the second stone revealed itself, having been hidden by a rocky outcrop. Both were very bulky, not like the slender slivers of the Stones of Stenness, or the Ring of Brodgar on Orkney, or Calanais on the Isle of Lewis, and they no doubt weighed many tons. The taller Brother towered over me at well over two and a half metres and the other was shorter, more stumpy. As if ignoring these majestic, dramatic ancient monuments to a forgotten faith, the farm track ran precisely between them before dipping downhill towards the Cow Stone. It too was very bulky, standing about three hundred metres to the north-east, looking as though it had been carefully placed in some sort of alignment with the stones up on the summit. But the tracks of modern traffic could not diminish the magic of this place.

Probably dragging them with ropes over log rollers, the people who erected the stones on Brotherstone Hill expended great labour to set them there. But while the details of their religious rites will almost certainly never be discovered, the reason why they raised the standing stones on the hill could not have been clearer to me. The panorama was so vast, so heart-piercingly beautiful in the midday sun, that to stand by the Brothers' Stones and simply gaze at it was thrilling.

Below us the valley of the Tweed, the great, life-giving river, widens as the Tweed winds its way east to the sea between the sheltering hills. To the south rise the foothills of the Cheviots, and I could see the rounded shape of Yeavering Bell, a place Cuthbert would come to know. Clouds clustered over the long hump of Cheviot itself and my eye was led to the watershed ridge running west, the line of the English border, where I could pick out the twin peaks of the Maiden Paps. Closer is the conical, volcanic shape of Ruberslaw and beyond it the hills of the Ettrick Forest. Much closer rise the Eildons, Roman Trimontium, the three hills that dominate the upper Tweed Valley: Eildon Hill North, Eildon Mid Hill and Eildon Wester Hill.

To the north-west, also close, is Earlston Black Hill, Dun Airchille, the Meeting Fort – all places where the strangeness and romance of Thomas of Ercildoune, True Thomas, Thomas the Rhymer, still clings. A prophet who foresaw the death of Alexander III in 1286 and whose predictions were widely believed and repeated, Thomas was perhaps the most famous Scotsman of the Middle Ages. The low ridges of the Lammermuirs lie to the north, many studded with groups of elegant white wind turbines, their sails turning lazily in the breeze. And standing up proud above the horizon in the north-east, I could easily make out the summits of Great Dirrington Law and Little Dirrington Law.

Forming a vast natural amphitheatre, the hills surround the courses of the great river and its tributaries on three sides. On the fourth, to the east, the horizon dips low as the patchwork of ripening fields edge down towards the seashore. With a hand on the cool mass of the taller of the two Brothers, its surface smoothed by the winds, rain and snows of 5,000 winters, I looked and looked out over a landscape I know intimately but one I had never gazed upon from this place. And somewhere in my memory, deep inside my sense of myself, the warmth of an old love began to glow once more.

I was born and raised in the valley of the great river and I know that my ancestors have been on the banks of the Tweed for many, many centuries, ploughing the fields, coaxing food from the soil, perhaps seventy generations living and dying within sight of the sheltering hills. The rich red and black earth is grained in my hands. But as I grew older, sat and passed (mostly) exams, heeded the advice of my parents 'to get on' and 'stick in at the school' and 'make something of yourself', I knew I would have to leave, have to tear up deep roots and go and live somewhere else. And so I did. After university at St Andrews, Edinburgh and London, I was, to my amazement, appointed to run the Edinburgh Festival Fringe in 1975 and despite my youth (I was twenty-five) and inexperience, I made a success of it. My big break; it led to everything else and a twenty-year career in television that brought rewards and occasional satisfaction. But I always nursed a need to come home, to return to the Tweed Valley, and having bought our little farmhouse, I decided to resign my well-paid job and begin the precarious life of a freelance writer. At forty-nine, I still had energy and ambition. And so for almost twenty years I have lived and worked in the Scottish Borders – and gradually taken all of its beauties and glories for granted.

When I climbed the hill to look at the stones, I had given little thought to the hill itself, the place the old peoples had chosen, but when I looked out from it, the elemental power of its beauty flooded back, an uncomplicated, unconditional love for the fields, forests, hills and intimate valleys of my native land, my home-place. That reminder was to be the first of Cuthbert's gifts.

And up on the hill, history was whispering. The Eildon Hills to the south-west rose abruptly from the floodplain of the Tweed, their flanks steep, their three-summit outline dramatic and distinctive. The name 'Eildon' derives from Old Welsh, the language spoken by native communities before Cuthbert and my Anglian ancestors brought early versions of English, and it means 'the old fortress'. It was attached to one summit, to Eildon Hill North, and it is closest to Brotherstone Hill. The name described a long perimeter of double ditching dug some time at the outset of the first millennium BC. But it cannot have been defensive. Probably topped by a palisade of wooden posts, the ditching runs for a mile around the crown of the hill, and instead of one or two gateways it has five. A huge force of defenders would have been needed to man these ramparts, especially the weak points at the gates, and if they had, they would soon have been thirsty. There is no spring or source of water other than what fell from the sky.

Eildon Hill North was not a fort but a temple, and a place of spiritual and temporal power. Inside the long perimeter, more than three hundred hut platforms have been found, enough to house a population of about three thousand. Lack of water and the need for all provisions to be lugged up the steep slopes probably meant a temporary occupation of these huts, almost certainly for significant points in the year; times of celebration, worship, perhaps sacrifice and perhaps propitiation of whatever gods were believed to govern the lives of mortals 3,000 years ago.

The oldest calendar in Britain revolves around the half-forgotten turning points in the farming year of the native Celtic peoples. It begins at the end of October with Samhuinn – in Gaelic: 'the end of summer'. It is now Christianised as Hallowe'en and seen as the beginning of winter and the moment when the clocks change. But the remnants of ancient, pagan practices still survive and versions of these would likely have been enacted by those who climbed Eildon Hill North on Samhuinn Eve.

Originally, guising at Hallowe'en did not involve outlandish dressing-up but a simple daubing of ash from the great bonfires that blazed in the darkness to effect a symbolic disguise, the cue for all sorts of licence between men and women. It persisted for millennia and in the 1796 *Statistical Account for Scotland*, kirk ministers in several parishes were complaining about 'A sort of secret society of Guisers made itself notorious in several of the neighbouring villages, men dressed as women, women dressed as men, dancing in a very unseemly way.'

There is persuasive evidence that belief in late prehistoric society attached great significance to the human head, perhaps as the repository of essence, of what might have been seen as the soul. Roman historians reported that European Celts were fond of collecting the heads of their enemies as grisly trophies, sometimes attaching them by the hair to their saddles, even preserving them in cedar oil. In Scotland, skulls have been found at several sites and their arrangement suggests that they were displayed in some way. As late as AD 70, the priests of Venutius, the native king of the Brigantes, a federation of Celtic kindreds on either side of the Pennines, set up a row of skulls to defend a rampart against the assault of Roman legions. This has been described as a ghost fence.

These ancient beliefs still sound a distant echo at Hallowe'en. When hollowed-out turnip or pumpkin lanterns have a candle placed inside and are set on a windowsill, it looks very much like a relic of paganism; a ghost fence still flickering in the early dark of winter.

At Imbolc in February, fires blazed once more inside the precinct on Eildon Hill North. The festival has been Christianised as St Bridget's Day and it was the moment of first fruits, when ewes began to lactate in anticipation of lambing. In early May, Beltane signalled the beginning of the time-worn journey of transhumance, when flocks and herds were moved up the hill trails to summer grazing. This was celebrated in Scotland as late as the early nineteenth century with ritual meals washed down with a liberal amount of alcohol. Not surprisingly, the kirk wagged its finger at what ministers knew to be a relict of paganism. The final nodal point came around in August. Lughnasa is now pronounced as 'Lammas' and a famous fair is held in St Andrews. This was the time grass-fattened beasts were bartered for slaughter or breeding, and the clustering of agricultural shows at that time in the calendar remembers the pivotal importance of the old Celtic feast.

All of these turning points revolved around the behaviour and needs of domesticated flocks and herds and the rhythms of a stock-rearing society. Weather mattered to our ancestors even more than it does for modern farmers. Three thousand years ago there were no large byres to overwinter cattle, there was little or no shelter from the winter's blast other than the folds of the land or the leafless trees of the wildwood. Severe snow, ice and cold, wind-driven rain saw many animals die and a parched summer could mean that they did not put on condition. To judge their prospects, farmers looked to the sky, and it seems that by the first millennium BC it was also where they looked for the gods that would help them rear their beasts and bring home their harvests safely.

Priest-kings climbed Eildon Hill North on the Celtic feast days to be nearer to their pantheon and to be seen by the divine beings who in some way controlled what happened on the Earth, who sent good and bad fortune, who sent snow, wind, rain or sun. They were sky gods, and the great enclosure on Eildon Hill North was not a fortress but a sky temple.

The Brothers' Stones do not stand on the highest part of their hill. To the west, there is a rocky outcrop, which hid the second stone as I approached up the farm track, and I sat down there to look at and think about the relationship of the great temple and the stones, if there was one. And how all of this might have been understood by Cuthbert.

The stones were probably dragged up Brotherstone Hill a long time before the ditches were dug on Eildon Hill North, perhaps a thousand or even two thousand years before. But I cannot think that this concentration of monuments was an accident. Perhaps the link was simple – the sky. Thunder, lightning and the drama of the weather have long been associated with the gods and their moods, and on Brotherstone Hill there is no better place to watch it change or settle, not even from the sky temple. Eildon Hill North does not have all-round panoramic views. Instead its longest vistas stretch to the east, to the twin stones and the Cow Stone at the foot of the slope. They felt and looked to me like a gateway, the doors of some sort of lost perception. I am very sceptical about all of the musings of those who follow ley-lines and other invisible trigonometry in the landscape, and I know I am straying perilous close, but there did seem to be some sort of alignment between the summit of Eildon Hill North, the Brothers' Stones and the Cow Stone.

Cuthbert cannot have been insensitive to the ghosts of the pagan past that swirled around the hill where he tended sheep. Both the Anonymous *Life* and Bede's speak of his missions of conversion, his attempts to defeat pagan belief in the Borders, in the hills around Melrose, persuading country people to turn away from amulets and incantations and back to the love of Christ. And yet the stones may still have been venerated in some way. Perhaps the young shepherd saw Brotherstone Hill as a Christian sky temple, a spiritual vantage point over the valley of the great river, the place where he could clearly see angels descend and ascend with the glowing orb of fire that was Aidan's soul. What modern eyes might have seen as an eclipse, a rare Blood Moon, where the sun turns its milky white surface to red, might have seemed like a vision to Cuthbert, metaphor rather than reality, the Earth and the celestial phenomena around it as themselves divine.

Turning over these half-formed notions about the sacred earth, the hallowed ground, a place where saints had walked, I sat for a long time by the old stones, watching cloud shadows chase across the sunlit fields before disappearing into the folds of the hills. Even though the account of Cuthbert's vision is infused with biblical tropes, from Jacob's dream of the stairway to heaven at Bethel in the Old Testament to the shepherds tending their flocks at night on the eve of the Nativity in the New, I was thinking about something beyond the scriptural. Below me, on the lower slopes of the hill, a flock of newly shorn ewes were bent over the sweet grass of fresh pasture, the lambs almost as big as their mothers. Thirteen centuries since the boy woke his fellow shepherds to tell them what he had seen in the sky, sheep were still grazing on Brotherstone Hill. And the skies above were still immense and dramatic.

When I ask myself a pressing question, more and more pressing on the threshold of my eighth decade, about what I believe, the provisional answer has nothing to do with the divine, a Christian god or any other, or indeed any conventional hope of an afterlife. I believe in an immortality of a different sort: the immortality of continuity, especially continuity in the same place. I will live on through the lives and memories of my children and their children. Even though that will fade in time, perhaps in only three generations, there will be a pile of dusty books somewhere with their forewards, dedications, blurbs and contents where some sense of my life and what I thought can be assembled. If anyone is interested.

With Cuthbert, even though there is no certain way to know his ancestral DNA, I feel that because mine is also Anglian, we have a direct connection. And on that warm and sunny day on the hill, I felt I could hear him on the edge of the wind, murmuring his prayers, looking up at what I was looking at, searching the sky for angels. But he could not have known that he was gazing upon a large part of what would become known as St Cuthbert's Land, a place where dozens of churches would be dedicated to him and where, for 1,000 years, documents would carry the phrase 'for St Cuthbert and for God'. And I did not know until that moment my journey had actually begun. I hoped the shepherd boy would lead me down the hill and that I could walk with his shadow to the monastery at Old Melrose.

3

In the Sacred Land

It had been a parched summer, the hottest and longest for more than forty years, with the sun beating down almost every day since the beginning of May. After a six-month winter, with the last snowfall in early April, the landscape was flooded with welcome light and warmth.

Temperatures climbed steadily and reached ninety degrees Fahrenheit in the last week of July. For weeks on end we sweltered in the high seventies and eighties, and virtually no rain fell. But then one evening the heat suddenly became very oppressive, and over the dark heads of the southern hills gunmetal-grey clouds quickly gathered. Far in the distance, thunder rumbled. By 9 p.m. it had grown very dark, and after a few spots it began to rain torrentially. While we ran around the farmhouse closing windows, there was a tremendous crack of thunder directly overhead that made us all flinch instinctively and sheet lightning lit the black sky.

An overnight downpour began, the raindrops huge, and it had a dramatic effect around the farmhouse, especially on the track. In towns and cities, heavy rain falls on hard surfaces – roofs, tarmac and pavements – and is usually taken away by efficient drainage. When it stops, and especially if the sun shines afterwards, it is as though it has never fallen. Here, heavy rain leaves its mark for many days. Much of the rammed earth and pebbles, compacted by vehicle wheels, hooves and feet, was washed away, exposing the stones of the old track, like Roman cobbles. It was worst down at the house, where the velocity of the flow was strongest. I could see that the downpour of the previous night had created a miniature water world, a network of tiny river channels, oxbows, deltas, lakes and dams, something that would delight a child. The bottom track is about a 15% gradient and 120 yards long, and so when the little rivers of rain reached the bend around the gable end of the old house, they had considerable force and carried a lot of pebbles and silt with them. As the camber forced the rainwater to turn, it cut out little river cliffs two or three inches high, as the fine silt piled up. It was like speeded-up land formation, millions of years of geology recreated by one night of torrential rain.

By the time I took the dogs out at 6.30 a.m., the sky had cleared and a watery sun blinked above the woods out to the east. As often after summer rain, the scents of the land were released: the musky smell of wet earth, the bitterness of leaves battered off twigs and branches by the huge raindrops, and the puddles on the tracks filled with the metallic whiff of silt and grit, which was dust only twelve hours before. More, less dramatic, rain was forecast for later in the day, but I decided to risk getting wet and to make a sustained start on the first few steps of my pilgrimage, my journey in Cuthbert's long shadow. His vision on Brotherstone Hill had deepened the young man's piety and ultimately persuaded him to go to Old Melrose to seek admission to the community, to become a novice monk.

Since I wanted to walk where he had walked, see something of what he had seen, I needed to work out the route of Cuthbert's journey from Wrangham to the monastery, what were probably the most fateful, life-changing steps he was ever to take. Always as a historian I have tried to imagine how people thought at the time and what prompted them to act as they did in the past, avoiding the attachment of modern motives and attitudes, thereby not falling into the trap of reading history backwards. Neither Robert Bruce nor Edward II of England knew how events at Bannockburn would fall out and the fact that we do should not colour the telling of the story of the battle. Nothing was inevitable.

So it was with all of the actors in our history. As he made his way between the hedges and looked out over the river valleys of his childhood, Cuthbert did not know where his decision to become a monk would take him, or indeed that the monastery would even admit him. Bede recounts something surprising. Instead of walking, showing some humility in imitation of Christ and the Apostles, Cuthbert rode high on his horse, carrying a spear, and with a servant walking beside him. This detail certainly marks him out as an aristocrat of some degree, but it is difficult to believe that he was armed on account of the countryside between Wrangham and Old Melrose being hostile, these fields and lanes he must have known so well. It seems much more likely he was showing off his status. Lack of self-confidence and uncertainty might have prompted him to think on all manner of possibilities: that his journey was a final taste of freedom, the last time he would look over the fields as a free man before he gave his life to God. And I am sure he wondered how strong his commitment was. Perhaps floating at the back of his mind was the possibility of rejection – either a change of mind on his part or a refusal of the monks to admit him.

This is a pivotal moment in Cuthbert's story and it is worth quoting the relevant passage at length from Bede's *Life*. His characteristically precise and crisp monastic Latin makes a difficult journey sound rather too smooth and inevitable, but Bede's business was more than historical. He wanted to establish the cult of St Cuthbert and no doubts or blemishes could be admitted in the story of the holy man's exemplary life. Bede was not so much reading history backwards as making sure it travelled in a straight line and in the right direction:

> Meanwhile the reverend servant of the Lord, having forsaken the things of the world, hastens to submit to monastic discipline, since he had been urged by the heavenly vision to seek the joys of eternal bliss and to endure temporal hunger and thirst for the Lord's sake as one who had been invited to the heavenly feasts. And though he knew that the church at Lindisfarne contained many holy men by whose learning and example he might be instructed, yet learning beforehand of the fame of the sublime virtues of the monk and priest Boisil, he preferred to seek Melrose. And by chance it happened that, having jumped down from his horse on reaching the monastery, and being about to enter the church to pray, he gave both his horse and the spear he was holding to a servant, for he had not yet put off his secular habit.

Ignoring modern roads and remembering that in the seventh century there were no bridges over the Tweed or the Leader (not since the Romans abandoned their great military depot at Trimontium, at the foot of the Eildon Hills), I had pored over large-scale Ordnance Survey maps, the excellent Pathfinder series in particular, looking for traces of disused tracks, and also consulted as many old maps as I could find. In order to reach Old Melrose, I had concluded that Cuthbert had had to go south-east from Wrangham to wade with his horse and servant across the Tweed at the Monksford, an ancient crossing about a mile south of the monastery.

From Brotherstone Farm, an old C road led to the hamlet of Bemersyde on a ridge above the river's floodplain. And, from there, I reckoned the route turned south and downhill to Dryburgh, where the romantic ruins of a twelfth-century abbey now stand. From there, Cuthbert would have followed the banks of the Tweed north to the ford. That made sense to me, as the most likely path. And when I was checking my maps one last time before packing them in my rucksack, I noticed that a solitary ash tree had been plotted on the Pathfinder by the side of the road, about four or five hundred yards east of Brotherstone Farm. I wondered if this was a relic of the row of ash trees pointed out by the Rev. W.L. Sime and the Berwickshire Naturalists' Club in the summer of 1930.

By the time I shouldered my pack (spare socks, pants, a towel, a waterproof, a copy of Colgrave's translation of the Anonymous *Life* and Bede's version, maps, a spare pen and notebook, a fully charged mobile phone, cheese sandwiches, chocolate and a hat) and gone off to look for the ash tree, the sun had climbed and my body warmer quickly became too warm. Despite walking back and forth on the road, consulting reference points on my two overlapping Pathfinders and searching amongst the thickly overgrown hedgerows and field-ends, I could find no sign of the ash tree, or even its stump, and retraced my steps to the farm and the C road, disappointed at a false start, something I had already experienced at Brotherstone.

Almost immediately, I knew that this was an ancient route. When the fraying tarmac swung west at Third Farm, the track grew narrow, made not for modern vehicles but for carts, riders and those on foot. On either side were heavily overgrown but deep ditches, dug to catch the rainwater run-off, and in the middle were the intermittent remains of a crest. Between the ditches and the fields ran long avenues of very old hardwood trees. Some had been blackened by lightning strikes, others had lost their heartwood and were regrowing around the edges, or had sent out suckers or seedlings. The hedges between the trees were broad and very dense, good winter shelter for small animals and birds, and full of summer and autumn goodness, with their harvest of rosehips, wild raspberries, brambles and haws of several sorts. It was impossible to see through the thick branches and abundant foliage.

From the farm at Third, the old road began to climb gently and through field entries I could see long, sunlit vistas to the south, the Tweed and Teviot valleys and beyond them, the watershed ridge of the Cheviots. I met no one on the road, and no farm traffic. The landscape seemed to doze in the sun and, as I began the metronomic rhythm of putting one foot in front of another, my senses began to drift, absorbing little more than the warmth, the scents of the land and its summer glories. On both sides barley fields stretched across undulating, free-draining ground, the ripening, heavy heads rippling in the breeze. Near the top of the rise I had been steadily climbing, the old road crossed a green loaning, a wide path that ran south to north. It had probably been used for driving flocks and herds to the high summer pasture on the flanks of the Bemersyde, Brotherstone and Redpath hills. Its hedges had not been trimmed for many years and the hot summer had seen them soar in height like rows of small trees. As I breathed the clean air, taking my time through this place of, it seemed, complete peace, I wondered why Cuthbert wanted to leave it for the seclusion and austerity of life at Old Melrose. Was there turmoil in his soul? Why was this not enough?

Later in his brief account of Cuthbert's journey to Old Melrose, Bede described his motivation in a single phrase, 'he preferred the monastery to the world'. With his servant walking beside him, his horse's reins in one hand and his spear in the other, the young man, perhaps only fifteen or sixteen years old, presented the perfect picture of the secular, even warlike world. In the Anonymous *Life* reference is made to Cuthbert at one time 'dwelling in camp with the army', probably having been conscripted into the royal Northumbrian host. But he was about to cast aside his spear, hand over his horse to his servant and leave the world 'for the yoke of bondservice to Christ'. The young man would soon pass from the familiar patchwork of fields, farms and villages and cross into the sacred land.

If I was right about Cuthbert's road to meet his destiny, it took him through the ancient hamlet of Bemersyde. No trace of the early medieval village is left, but the origins of its name point to real antiquity. It means something like 'the hillside where bitterns call', birds that have been described as the Trumpeters of Bemersyde. To the south-east, in a dip between two ridges, lies Bemersyde Moss, a small loch surrounded by tussocky wetland, the perfect habitat for these fishermen, relatives of the heron. The place-name is an example of transference. In Old English, the language of Cuthbert, *'bemere'* means 'to trumpet'; its call, instead of the bird itself, found its way onto the map. It is likely that the hamlet existed in the seventh century during the Anglian takeover of the Tweed Valley, and after I had walked out of the tree-shaded lane, the C road I had been following joined a B road bathed in bright sunshine, where its houses clustered.

Most of these are modern, strung out on either side of the road that runs east to west, and their views must be sweeping. Around each are arranged neat and colourful gardens, the deposit of much care and effort. As I passed, a dog leaped up and barked suddenly at a gate and the owner came to calm the old collie and apologise. There was no need, but what struck me was the sense of contentment the lady seemed to have as she talked about the remarkable summer and I praised her beautiful garden of roses, geraniums and hydrangeas and the pots, baskets and borders of rich colour. As I walked off down the road, I wondered about my own choices in life. A comfortable modern house with long views, a decent pension, plenty of time to indulge interests and no need to keep striving, pushing and hoping – all of that suddenly seemed very attractive, something my wife Lindsay and I could easily have opted for.

Instead, of course, we took the harder and, more than occasionally, I think, the dafter path. The joys have mostly outweighed the difficulties, but the life of a freelance and its income is precarious and has significantly diminished over time. More than once, we have had to scrape through a tough year. And at sixty-eight, I worry that my earning capacity is beginning to slacken. More and more often, I look at bungalows like those at Bemersyde and wonder about comfortable sofas, lie-ins on a weekday, playing at writing something I don't need to finish because I don't need the money, and joining a book group. But so far these thoughts are fleeting, driven out by the likelihood that I would soon become bored, vastly overweight and possibly over-fond of New Zealand sauvignon blanc, if I am not already. For me at least, for the moment, hard work, no holidays, only the occasional day off and bouts of intermittent anxiety are probably healthy.

And probably inevitable. Just as Cuthbert may have been surrendering to his essential nature, recognising that his piety, his wish to leave the world and strive to know the mind of God, would always prevail over his inherited status as an aristocrat of some sort, so I have begun to realise that I could have lived no other sort of life. I am a risk-taker, someone who sees how well things can turn out and never seriously considers how badly wrong they can go, or if that sounds a little melodramatic, someone who could not bear to be too safe and take refuge in the false securities and certainties of routine. I could not have stayed in a salaried job, served my time until my pension had fattened up, for I would have lost my life if I had wasted all those years just turning up every Monday morning, ticking off the months and years until retirement, wishing my life away.

Instead, I took the risk of depending on myself, not leaning on the support of an institution, and trying to make a living from what was in my head. I don't much care if that sounds self-aggrandising or even puffed-up; it is nothing less than the truth. I had to be true to my nature – although that sounds a little pat and premeditated. At the time it didn't feel like that, it was just something I had to do. Even though I had not properly thought out the consequences of diving into the depths of uncertainty twenty years ago, I knew I had to get out of corporate life and accept all the insecurities that came with that decision. And even though the life of a freelance is very dependent on the wishes, whims and appetites of others, those who commission our work or grant money to support it, I suppose I am content with the conditional truth that I have at least been my own man.

But alternative choices can sometimes be surprising and illuminating. At the end of 2017 I met a man I had not seen for fifty years. We played representative schoolboy rugby together in the Borders before he went off to Edinburgh University to take a degree in geology and then a job with De Beer in South Africa. Very dashing, but certainly uncomfortable and almost certainly risky, as well as a radically different choice from mine to stay and make a life in Scotland. When we met for supper in Melrose, I worried that we would have little or nothing to say to each other. In fact, it turned out to be fascinating.

Alone for weeks on end in the African bush, taking sample cores, looking for likely places where diamonds might be mined, my old rugby-playing friend often found that he had spoken to no one for long periods. Instead, he began to read the novels of Walter Scott in publication sequence, and not only worthwhile in itself, it was a habit that reconnected my friend with the Borders. The famous Scott quote, 'Breathes there the man, with soul so dead, Who never to himself hath said, This is my own, my native land!' was precisely apposite. My friend's mother is still alive and he flies north from his home on the coast near Cape Town to the Scottish Borders at least two or three times a year. His connection to our home-place is powerful. As is mine. When I resigned my job, there was no other choice I could consider except to come home, to return to the sacred land.

Continuity and attachment to place are rarely more evident than in the story of Bemersyde House. Turning downhill from the hamlet, I passed its gates and saw the sixteenth-century tower house in the distance. The same family has lived there for eight centuries since it came into the possession of a Norman warrior called Petrus de Haga. By the late thirteenth and early fourteenth centuries, the name had changed slightly but become so established that Thomas the Rhymer could recite:

Tyde what may,
Whate'er betyde
Haig shall be Haig of Bemersyde.

And so it has come to pass. Alexander, the grandson of Field
Marshal Earl Haig, the commander of the British army on
the Western Front during the First World War, now lives in
the old house.

Below the gates, the road begins to run down gently off
the ridge and becomes deep and heavily shaded, a sign of
real antiquity, since it has sunk far below the level of the
fields beside it simply because of the centuries of feet, hooves
and wheels, the tread of people, horses and the ruts made
by carts. Powerful knots of tree roots cling like octopuses
to the high banks and their thick foliage shades the road for
a few hundred yards. It eventually leads to an oddity, a disso-
nance.

When the sixteenth century and the Reformation signalled
the end for Scotland's monastic communities, abbeys often
fell under the control of commendators, usually aristocrats
who eventually appropriated their lands. The holdings and
power of many aristocratic families are built on the patrimony of the church. The Erskine family, earls of Buchan,
were given Dryburgh Abbey and much of the nearby prop-
erty that had been gifted to the monks over many years. By
the early nineteenth century, David Stuart Erskine, the 11th
earl, had become fascinated, even obsessed by Scotland's
history and heroes. He founded the Society of Antiquaries,
whose collections formed the basis of those of the national
museums of Scotland. Walter Scott knew Erskine well and
was uncharacteristically ungenerous, saying that he was a
man whose 'immense vanity obscured, or rather eclipsed,
very considerable talents'. On that sunny day below

Bemersyde, it seemed to me that an understanding of what made good art was not amongst them.

The deepened lane leads downhill to a brown sign on the left that points to 'Wallace Statue'. At the end of a winding wooded track, perched on a high, precipitous bank, is a monstrous red sandstone sculpture of one of Scotland's great heroes. William Wallace, the victor of Stirling Bridge in 1298, stands twenty-one feet high on a ten-foot plinth, staring sightlessly westwards over the Tweed Valley. Made by a local sculptor, John Smith of Darnick, no doubt to a precise prescription from the Earl of Buchan, it could never be mistaken for the work of Michelangelo. In fact it is profoundly ugly. Holding a broadsword almost as tall as himself in one hand and a shield bearing the device of the Saltire in the other, the hero looks more than a little gormless, a slightly puzzled expression above his bushy beard, as though he were lost. On his head, and therefore difficult to make out from thirty-one feet below, a version of an iron helmet that owes more to the Wehrmacht than anything medieval has some sort of winged creature attached. Perhaps a pigeon. Mercifully, the hardwood trees around this thoroughly duff object have begun to hide it. From a distance, to the west, all that can be made out is the head – until autumn sheds the friendly leaves.

Further down the road to Dryburgh Abbey I came across something much more eloquent, a fragment of cut stone that was definitely more pleasing, clear confirmation that people had walked and ridden this way for very many centuries. Easy to miss in the left-hand verge, hard against the edge of the tarmac, sits a cross socket. Now filled with rainwater, it once held a tall and impressive cross that offered a place to pray and that marked a boundary. Abbeys, priories, convents, churches and other places of pilgrimage often lay inside a wide precinct whose outer limits were fringed by crosses set up by the sides of the roads that led to the sacred sites. Around Coldingham Priory on the Berwickshire coast, founded in the mid-seventh century and a place Cuthbert visited, there were at least three crosses at three approaches: Applincross, Whitecross and Cairncross. No more than this unconsidered little stump survives from Dryburgh, but there must have been others. When Cuthbert rode past it, he left the temporal world and entered holy ground.

Before a long avenue of trees wrapped the road in shadow, wide vistas to the south and west opened over the rich, pale yellow of ripening barley. Rain was gathering in the west, a grey veiling drifting across the hills of the Ettrick Forest and the shelter of the deep lane made it difficult to judge where the wind blew. I quickened my pace downhill to Dryburgh, noticing that gaps in the trees to my right showed what an ever-present landmark Eildon Hill North was. It seemed to be following me around the landscape.

Dryburgh is the only one of the four Border abbeys not found in a town and consequently its fabric is more complete, having suffered much less at the hands of stone robbers. After the Reformation, the masonry of the great churches at Kelso, Melrose and Jedburgh could be seen in the walls, and no doubt the foundations, of nearby houses. Sheltered by stands of ancient trees – one yew is said to have been planted by the monks in the early twelfth century – and surrounded by a vallum, a ditch deepened in modern times to keep out grazing and browsing animals, Dryburgh Abbey is a very romantic ruin, and also strange, other-worldly.

Many years ago I made a television series called *The Sea Kingdoms*, the story of Celtic Britain and Ireland. We filmed in Cornwall, Ireland, the Western Isles and in Wales. On our way out to St David's in Pembrokeshire, surely the only cathedral village in Europe, we took a detour to a place associated with another early holy man, St Justinian. According to the map, near the coastal hamlet was an interesting series of prehistoric remains, a small stone circle and a dolmen, a megalithic tomb of two upright stones that supported a horizontal capstone. I thought some shots of these against the sea might be useful in opening credits.

The landscape was patterned by a warren of narrow paths threaded between rocky outcrops and tall clumps of impenetrable gorse. It was easy to become disoriented. Near the dolmen, we came across an unexpected small cottage with a fenced front garden full of colourful wooden objects: stripy posts, a large doll and a cart. Bunting fluttered on some of them. As our camera and sound men gratefully put down their kit, my director and I knocked on the door. Perhaps we were trespassing and needed permission to film. There was no answer at first, and then I began to make out a low drone, almost like a growl but not something that sounded like a guard dog. When there was no answer and we retreated, it was replaced by a high-pitched whine, a keening that slowly built up but did not seem to come from the cottage.

After exchanging glances, we walked back to the dolmen, wondering what we might meet around each corner of the path, shot some general views, packed up our kit and walked, quite quickly, back to the cars. More than strangeness, this little enclave in the landscape had a powerful atmosphere, something malign, and even though it was a bright day, good for filming, we all felt uncomfortable and were relieved to park at St David's and set up our next sequence of shots amongst the crowds.

Dryburgh's strangeness is not malign, it seems to me, but

it is not a place of settled peace either. For some who visit, God may be close, but I had a powerful sense of other presences, perhaps the spirits of the pagan past were flitting in the shadows of the old trees. The early history of the abbey and its site is scant but it might cast a dim light on these competing impressions.

All of the original locations of the great twelfth-century monastic foundations in the Borders are to be found in the loops of rivers. Jedburgh Abbey rises above a sinuous bend in the Jed Water while Dryburgh, Kelso and Old Melrose are all bounded in part by loops in the course of the Tweed. Such a wide and deep river was a real barrier for millennia, until the bridge-building of the modern age, and it offered a degree of seclusion, a clear division between the temporal and the sacred worlds.

Early Christians were attracted to sites like these because they were impressed and inspired by a group of ascetics known as the Desert Fathers. Perhaps the best known in sixth- and seventh-century Western Europe was St Anthony of Egypt. Copies of his *Life*, written by St Athanasius of Alexandria, found their way to Britain and Ireland, and scholars believe that both Bede and the anonymous biographer of Cuthbert were able to consult this short but remarkable text. It relates how in the late third century Anthony was raised in a Christian household of some considerable means. When both his parents died, he seems to have had an epiphany. Giving away all of his inherited wealth, a farm of 300 acres and much else of value, he put his younger sister into a convent and embarked on a life of piety and asceticism. After many battles with the Devil and his legions of demons, he sought a solitary life, shutting himself up in a tomb at one point, starving himself and forcing himself into cycles of prayer and vigil. Although Athanasius nowhere states this explicitly, there is a sense threading through the narrative (and others) of the hermit using mortification of the flesh to induce a trance-like state so that he might have out-of-body experiences, something that may have seemed to detach his immortal soul from his worldly flesh. This appears to have been a continuing practice amongst these early saints.

Eventually, Anthony fled to the deserts beyond the fertile valley of the Nile to pursue a life of solitude that might draw him nearer to God. But his fame had spread and, according to Athanasius, 'cells arose in the mountains, and the desert was colonised by monks who came forth from their own people . . . and he directed them like a father'. And then Anthony retreated even further, to a holy mountain, his 'inner mountain', where he lived out the rest of his harsh life. Surviving the persecutions of the Emperor Maximinus II in 311, he died some time later after giving instructions that his body was to be buried in secret so that it could not become a focus for reverence or pilgrimage, something else that would be clearly echoed at the end of Cuthbert's own life.

The teachings and habits of the Desert Fathers inspired early Irish monks to seek places apart from the world, places of solitude with no worldly distractions. In a Gaelic rendering of *deserta*, they called their hermitages and communities *diseartan* and it is a description that survives in the Scottish place-name of Dysart on the Fife coast, Dyzard in Cornwall, several Dyserths in Wales, and many Diserts in Ireland. Celtic monks let water, both salt and fresh, take the place of the sands of the Egyptian and Judaean deserts, and in the loops of the River Tweed at least two and perhaps four disearts were established in the sixth century as the spirit of Anthony reached across a continent.

Some time around 522, before Columba sailed to Iona in 563, an Irish monk called Modan came to Dryburgh, where he appears to have settled for a time. The son of an Irish sub-king, he has left shadowy traces of his journeys around Scotland. Ardchattan near Loch Etive was once Balmodhan, the settlement of Modan, Kilmodan on the Isle of Bute was his church, St Modan's High School in Stirling remembers his presence in the Forth Valley, and his relics are said to have been enshrined at St Modan's Church at Rosneath, across the Gare Loch from Helensburgh. In September 2015, local archaeologists rediscovered St Modan's Well in the woods above Glendaruel in Argyll and found pebbles of bright quartz around it that had been left by pilgrims who had come to pray there and seek his blessing over many centuries before the Reformation.

At Dryburgh nothing remains of Modan's presence except for the faint echo of a story. Having reluctantly taken up the office of abbot, he resigned and left the monastery so that he could be free to follow the life of a hermit over in the west of Scotland, near Dumbarton. Given that his relics found their way to Rosneath, it is reasonable to suppose that the Irishman died there. All of these references and locations suggest that the widely travelled monk was a real pre-Columban presence in the Borders, even though no material or documentary record of him can be found.

When Cuthbert rode past the cross by the road and made his way with his servant downhill to Dryburgh, he may have known the story of Modan, but the fact that he moved on to Old Melrose probably means that the early foundation had not survived the saint's departure.

But I think that Modan's spirit still flits around the ruined transept, the presbytery beyond it, the cloister and the tumbled walls of the twelfth-century abbey. It was probably built on that site because he had chosen it, dedicated it to God, made it sacred with prayer, and the stories lingered as the centuries passed. And more than that, I believe that like many Irish monks Modan made a link with the pagan past. He will certainly have worn the Druidic tonsure. Monks influenced by the doctrines of the Church of Rome had their hair cut from the crowns of their heads, leaving a fringe around the temples and at the back. This was done in imitation of Christ's crown of thorns. By contrast, Irish monks and many in Scotland were tonsured across the crowns of their heads, from ear to ear, giving them very high foreheads. Scholars believe that the pagan priests of the first millennium BC, the Druids encountered by the Roman invaders, wore their hair in the same way, their foreheads probably marked by tattoos.

On their missions of conversion of the sixth and seventh centuries, Christian priests and monks were mindful of the advice of Pope Gregory the Great. In the 590s, he ordained that instead of destroying pagan sites and temples 'they should be sprinkled with holy water' and used as places of Christian worship. The pagan festivals of Imbolc, Beltane, Lughnasa and Samhuinn should be celebrated not in veneration of idols but for the sake of 'good fellowship'. This pragmatic approach recognised that for many conversion was not a blinding light, an epiphany or a moment when the whole world changed, but a process whereby beliefs held over millennia gradually shifted. And it was sensible to worship a Christian god at places that were already sacred. They retained their sense of spirituality and also people knew where they were.

For these reasons, I believe that many sites whose history appears to have started only with the coming of missionaries were in fact sacred long before they arrived and that their original sanctity only withered slowly. Like St Anthony, Modan will have been an ascetic, practising mortification of the flesh in many ways. And in itself, I think that this too was a vital link with the beliefs of those who climbed Eildon Hill North four times a year to light fires and celebrate the turning points of the farming year. Unlike the modern Christian God of love and forgiveness, the pantheon of deities who governed the lives and fates of the peoples of Celtic Britain and Ireland, and also those Angles, Saxons and others who sailed the North Sea in search of land, needed to be propitiated. That is, the malignant, aggressive nature of some of these pagan gods had to be neutralised by acts of sacrifice, and often blood needed to be shed. The discovery of what are known as the bog bodies, prehistoric corpses sufficiently well preserved to show the marks of ritual killing, shows that human sacrifices were sometimes required if the pagan gods were to be persuaded not to send thunder, wind and rain to wreck the harvest, or a pestilence to visit the land. Fear was as powerful as faith.

The notion of human sacrifice was not confined to pagan cultures. The early Christian God of the Old Testament asked Abraham to sacrifice Isaac before relenting, but he could also be vengeful, and for many centuries the awful prospect of the fires and pits of Hell, or eternal torment, was very real indeed for most believers. It seems to me that the ascetic monks and hermits were also practising a version of propitiation, a way of pleasing and appeasing God, committing painful acts not only in pursuit of personal purity, piety and the transcendence of the flesh but also undertaken on behalf of communities. For that latter reason, saints were readily venerated, thanked and seen as figures who floated in a sphere between the mortal and the divine and who could intercede with God on behalf of many people. It is thought that the wide proliferation of early saints (each village in Cornwall seems to have its own) was a gradual substitution for local gods.

Like many of his contemporaries, Modan will have embodied much of this continuity and I think he was drawn to Dryburgh because it was already a holy place for the pre-Christian communities who lived along the banks of the Tweed. And that history has not entirely fled. It can be sensed, buried deep on the site of the twelfth-century abbey, where enough of the fabric survives to allow a good reconstruction of monastic life: the bells ringing the canonical hours at dead of night, the chant of the psalms, readings in the chapter house, the whispering as monks gossiped behind their hands, and all the bustle of a busy community going about the business of worship, management and food production. But under the daily din, the silences as the monks shuffled down the night stair to the church for lauds and other intervals of peace, another river runs, one whose course can only be intuited.

I am certain that Walter Scott knew and felt that Dryburgh was different. Of course with his love of romance, he will have felt at home in the old medieval abbey, its scheming abbots, priors, cellarers and novices gossiping, taking part in medieval politics, celebrating the drama of the Latin mass in the presbytery. But in his life Scott was aware, I think, of a strangeness in his own nature, a sensitivity to another world beyond history. In *Ivanhoe* and others in his vast canon of historical novels, Scott showed a powerful longing for the past that sometimes glows with roseate nostalgia but also connects with the ethereal, the uncatchable, something fluttering on the edge of his imagination. And few novelists have ever captured a pungent sense of place in the long past so well – but sometimes not completely. There is often something of the unexplained in Scott's stories. Perhaps that was the deeper reason why he chose to be buried at Dryburgh, not just because it is beautiful but because its roots in the long past reached far down into the earth of the river peninsula and were unknowable. Scott knew that ghosts flitted amongst the ruins.

Next to Scott's solid granite sarcophagus stands the small, simple headstone of a soldier: Douglas Haig, Field Marshal Earl Haig. He wanted the same memorial that was set up to those hundreds of thousands of soldiers who died under his command in Flanders and he is buried at Dryburgh simply because it is close to Bemersyde. As I stood by these two graves in the north transept, it struck me that there was not an obvious and extreme contrast but a clear connection between the unwilling architect of great slaughter on the Western Front and the man whose stories went a long way to inventing the Scotland that many regiments believed they were fighting to protect.

I spent less time at Dryburgh than I expected, perhaps because I was anxious to catch up with Cuthbert and be on my way. In the wooden hut that serves as a ticket office and small shop, I bought a packet of fudge to keep me going on my journey to Old Melrose. I would need it.

* * *

The road from Dryburgh turns sharply downhill to the flood plain of the Tweed and past some trim new houses built not from bricks or breezeblock but a good deal of cut stone. I passed some beautifully squared blocks of red and yellow sandstone sitting on pallets by the roadside. The starkness of these new builds will weather down well as the sandstone mellows with the years and their gardens grow up. I noted high walls and an electric gate around one house and supposed that rural crime penetrates everywhere, even down this half-hidden old road. I passed a footbridge over the Tweed but ignored it. A mile further north was Monksford and I planned a baptism of sorts, to wade across just as Cuthbert and his servant would have done, leading his horse behind them. After the driest summer for forty years, the river should be low.

At the foot of the road, on a small mound that looked artificial to me, stood another imposition from the eccentric Earl of Buchan. His fascination with Greek mythology drew David Erskine into episodes of laughable daftness. Apparently he once held a soirée where he dressed up as Apollo on Mount Parnassus with nine young ladies dancing around him as his muses. Goodness knows what they made of it. On this mound in front of me, at the beginning of a long, straight stretch known as the Monks' Road, he had masons build a small, circular, pillared structure he called the Temple of the Muses, and in the centre a statue of Apollo was installed. It has mercifully gone now and been replaced by a modern bronze by Siobhan O'Hehir of four naked women facing in four compass directions to represent the seasons. They look well, their poses not frozen but somehow sinuous and liquid. By contrast, I am certain that the original statue of Apollo would have borne a distinct resemblance to the cavorting earl. Littering the landscape in this way, plonking inappropriate objects in it for his own aggrandisement and amusement, is more than irritating. Just because they are relatively old, it should not automatically mean that this ridiculous temple and the monstrous statue of William Wallace should not be demolished. Far better to remember the monks who walked the old road beside these oddities: Modan, Boisil, Cuthbert and other later figures whose stories were of this place, men who searched the big skies above the river for signs of God's presence and who helped make the land look as it does now.

Beyond the mound, the long straight track of the Monks' Road headed north. The Pathfinder map showed that it led to Monksford, and my determination to splash across the Tweed in Cuthbert's wake. Supplying access to a smart wooden hut on the banks of the river, one used by those paying handsomely to fish the pools of the Dryburgh Upper beat, the road was in good repair and I made good time. At the end I could see a much more overgrown track beyond a metal gate. Tree-lined and curving down to the riverbank, it seemed not to be much used.

In my rucksack, I had packed spare socks, pants and a towel. My plan was to stuff my boots, trousers and socks in the pack and towel myself dry after I had crossed the monks' ford and gained the farther bank. About halfway down the track, I spotted a likely, leggy rowan and, with my penknife, cut a staff from it. As I waded the river and its uneven bed of smoothed stones, I wanted something to steady me and test the depth in front. Shaded and quiet, I noticed that the track had been used recently when I saw the marks of horse-shoes in damp places, and as I finished the last of my fudge I wondered how recently someone had ridden here. I was certain I was following Cuthbert on his horse, for the track was old and there appeared to be no other way down to the ford he must have used to reach Old Melrose.

However, puzzlement and disappointment waited once more. Instead of leading to the riverbank, the track simply petered out amongst some broadleaf woodland. Beyond it the banks of the Tweed were overgrown with bushes, nettles and the wide, rhubarb-like leaves of hogweed. I thought I could make out the shallows of the beginning of the ford but no path led to it. And off to my left I could see where the rider had gone. There was another track leading along the river, but it went back the way I had come. It was probably used by fishermen.

Only a little daunted, and of course rationalising that the path must have disappeared long ago since no one had been daft enough to use the ford for at least a couple of centuries, I used my rowan pole to thrash aside the hogweed and also warn me of sudden, unseen, ankle-twisting ditches in the overgrown bank. Of which there were several.

On what seemed like an area of level ground (but not free of nettles), I checked to see that no one was around to watch the crazy person strip to his pants and shirt, then stuffed everything into my pack. When I splashed at last into the river, a family of ducks erupted a little way upstream and the sun came out to glint brilliantly off the water. It was cold but not icy, and I could see where large flat stones had been laid near the bank, clear remains of the old ford. Some were the same colour of red sandstone that had built the Temple of the Muses.

I had reckoned the Tweed was about sixty or more yards wide at Monksford, but after I had carefully gone about twenty-five yards and the water was up to my knees, I found it difficult to see the bottom and judge what was in front of me. The channel by the far bank was in shadow and I had no idea how deep it might be. The surface was flowing smoothly, like a large volume of water, no stone or shallows broke it and the river seemed to be moving much more slowly. The small, wing-like shapes of sycamore seeds were eddying, not flowing directly downstream. Prodding with the rowan pole, I suddenly felt it go down much further and nearly lost my balance. My heavier pack didn't help and I rocked a little. I reckoned the pole touched bottom at about three feet, waist height. So, no. There seemed to be an invisible shelf rather than a gradual incline. So, not that way.

I turned upstream, thinking I might have lost the direction of the ford. Even though the maps all showed it running the shortest distance, directly from bank to opposite bank, perhaps it crossed on the diagonal? No. Not that way either. The far channel seemed deep there too. My baptism into the world of Cuthbert was baulked by a real obstacle.

After more splashing around, almost capsizing again after slipping on some big rounded stones, I decided to go back, to give up, very reluctantly. Without really articulating it beforehand (all I wanted to do was follow Cuthbert as closely as possible), I suppose I saw the wading of the river as a form of informal baptism into his world. But like bridges, fords need to be maintained. The spates of centuries of winters had probably shifted what shallow footing there had been, as the great river reclaimed its natural course. In winter, crossing must have been impossible, as rain and snow swelled the current. Boats would have been the only option. For me there was nothing for it but to be sensible, wade back to the bank to dry off, put my jeans, socks and boots back on, and retrace my steps back to the Temple of the Muses and the footbridge below it.

I had barely begun this journey with Cuthbert but had already seen several reverses and false starts. But in one sense at least I felt I had been close to him. On the road from Brotherstone, I had met no one. On a sunny July morning I had enjoyed a few hours of real peace as I moved through the summer landscape, something I would come to think of as the peace of Cuthbert.

Recently refurbished, the footbridge looked splendid. The views up and downstream revealed broad areas of white, dried-out river boulders and stones below the overgrown banks, but even though the water was low the channel under the bridge seemed deep.

After I had crossed, I came upon a small car park. Signs told me this was a section of St Cuthbert's Way. Beginning at Melrose Abbey, then climbing the saddle between Eildon Hill North and Eildon Mid Hill, descending to Newtown St Boswells before going on to the village of Bowden, it is not a way Cuthbert would ever have come. But it does pass some beautiful views and, having climbed the long steps to the bank above the river, I came across a steady stream of walkers who were enjoying it, stopping often to take photographs on their phones. One bench had been set up to look north, and of course Eildon Hill North dominated the landscape.

After a long climb up a winding wooden stair, the path snaked through dense woodland above the river and footbridges had been built to cross the deeper wooded deans. After about half a mile, I saw that a recent gale labelled Storm Hector (why has this childish American habit of anthropomorphising destructive weather been adopted?) had blown down a big ash tree and it had landed squarely on a bench, pulverising it. There seemed to be a message there. The path occasionally forked and I found myself following it away from the river. Around a corner, I was suddenly assailed by the roar of traffic above me, a bridge carrying the trucks and cars of the busy A68. This riverside woodland suddenly felt like an underworld, parts of a palimpsest, layers below the thunder of the twenty-first century. A helicopter flew low, unseen, and it seemed to make the trees vibrate. The peace of Cuthbert was shattered and once more I decided to retrace my steps to look for a path by the riverbank.

When at last I emerged from the green shade of the woodland, I found I had been going not so much in circles but back and forth, so that I had only walked half a mile in half an hour, poor progress. The path by the Tweed had almost been overwhelmed by the broad leaves of hogweed, and when I eventually reached the bank opposite the ford it was very difficult to get close to the water to see how deep it was. I met a fisherman in chest-high waders who told me I had been wise to turn back. Not only was there no one about if I had got into difficulties, I had been standing on the edge of a salmon pool he had only ever thought it safe to fish from the bank or from the place where I had been splashing around in the shallows of the far side.

Some wooden signs and two inexplicable red and blue arrows suggested at least three directions of travel, but I wanted to stay in sight of the Tweed. Someone had taken the trouble to strim the long grass and nettles to open a track and I assumed that it led to Old Melrose. But after a few hundred yards it simply stopped dead on the edge of another deep pool. By this time, I could see the high bank where the river had turned to make its loop around Old Melrose. The site of the old monastery was close and after several reverses and many steps retraced, I did not want to turn back yet again and look for another path. The problem was the alternative – a climb up a high bank to my left. The map showed the woods running out into fields at the top; somewhere up there a track might be found.

There seemed to be plenty of saplings growing out of the slope, most of them as thick as my arm, and for about twenty feet I made good and careful progress. But then I had to swing across the face of the bank for a good handhold on a sapling that would lead me up what looked an easier route. But when I slipped on the damp, peaty earth I cannoned heavily into the little tree, hitting it with my chest. Unfortunately a hard plastic buckle on my rucksack was between me and the tree and I heard the unmistakable click of a rib snapping.

I knew that sound and what it felt like. Some years ago, when I had some money, I bought myself a splendid dapple-grey gelding. Standing almost seventeen hands, Sooty (daft name, I know, but he had black legs) could easily carry me and we enjoyed some memorable hacks in the countryside around our farm, and with a professional rider aboard he won the Novice Working Hunter class at the Ettrick and Yarrow Show. But one morning, in the arena next to the stables, I rode him without stirrups. Absolutely daft for such a poor rider. Instinctively I held on with my legs, squeezing him around the girth and, well schooled as he was, that sent him off into an abrupt canter and me flying out the side door. I fell hard and broke several ribs on my right-hand side. Exactly where the plastic buckle, me and the tree had collided. It hurt, but not so badly that I couldn't make it to the top of the bank, and from there to the plateau of the river peninsula of Old Melrose. I wondered if repeated and inadvertent mortification of the flesh ranked alongside the self-inflicted torment practised by the monks who had crossed the ford and trodden the path to the diseart thirteen or more centuries before I made my undignified entrance.

Even more than the loop of the Tweed at Dryburgh, the river almost makes the peninsula an island at Old Melrose. Close to where I had scrambled up to the top of the bank, its course pinches tight, not quite joining, before it is pushed around 280 degrees by a massive river cliff gouged out by the glaciers of the last ice age. Near-vertical in places, it towers above the site of the monastery and adds to a powerful sense of enclosure. The topography of Old Melrose is surprising. Much of it is a high plateau that looks down on the river and affords a long southern vista to Monksford and Dryburgh; immediately below it is the fringe of a broad, grassy floodplain. Incongruously, there was a large canvas tipi pitched on it the day I arrived.

Now heavily wooded, the peninsula is part of a well-managed estate dominated by Old Melrose House. It stands on the highest point of the plateau, close to where the early medieval chapel of St Cuthbert was built, and around it are grass parks grazed by sheep and one or two horses. The buildings of the old dairy farm are bounded by woods and bright barley fields, early to ripen in this hot and sunny summer. Most of its byres are now converted into a tea room, a bookshop and an antiques shop. On the warm afternoon when I mortified one of my ribs, the discomfort was much eased by this beautiful, sylvan scene, a peaceful place that gave no hint of its ancient, austere existence.

4

Soul-Friends

Names tell stories, especially place-names, and sometimes they even move, taking their history with them. Old Melrose is so called because a New Melrose came into being, a consequence of dynastic politics. Scotland's most modernising medieval king, a man who made a decisive break with the Celtic, Gaelic-speaking past, was David I. Raised at the court of the Norman-French Henry I, he was the sixth son of Malcolm III Canmore and his prospects of succeeding to the throne of Scotland were unlikely. But he had talent, and fortune fell out happily for him. Fluent in French, schooled in Western European culture and steeped in the dynamics of politics, he was, as contemporary writers consistently assert, a most perfect knight. He was also devout and interested in the new, reformed orders of monks who were becoming influential in France. When he became an earl with wide lands in Southern Scotland, he invited communities of these men to found monasteries in the Tweed Valley.

In 1124 Alexander I died without an heir and his younger brother, Earl David, succeeded. Anxious to strengthen the bounds of his new kingdom and to retain control of the Church, a key engine of royal government, the king had a problem with Old Melrose. With the cult of Cuthbert securely anchored in England at Durham, where the spectacular new cathedral had risen on the river peninsula of the Wear, many of the sites associated with the saint were claimed by the prince-bishops, including Old Melrose. But David wanted to re-found the monastery there and he had invited a group from the order of Cistercians to come to the Tweed Valley. To appease Durham, he exchanged St Mary's Church in Berwick-upon-Tweed for the ancient site – and immediately ran into another problem. For reasons now lost, Abbot Richard and his Cistercians refused to build on the river peninsula, even though it was the holiest and most famous church in the Borders, a place where saints had walked and worshipped. They preferred to found their new monastery at a place called Little Fordell, two and a half miles to the west, beyond the ruins of the Roman depot at Trimontium, on the southern bank of the Tweed. But probably at King David's insistence, they kept the name of Melrose.

Folk toponymy associates it with the mels or the mallets used by masons to build the abbey and the Rose Window at the east end of the church. But in fact it is a synthesis of two Gaelic words. *Rhos*, or Ross in the anglicised spelling, means 'a promontory' and describes the old river peninsula, but *maol* is more obscure. It can mean 'bare' and many believe that the definition ends there. 'Bare promontory' was how Old Melrose looked in the seventh and eighth centuries, long before the mighty trees of the nineteenth-century estate were planted. But, in fact, that is a misreading, and the reality is more interesting. *Maol* can also mean 'bald', and the old Druidic tonsure cut across the crown of the head made Celtic monks look bald. And from that, it acquired a further meaning that can be detected in the Christian name of Malcolm. It comes from *Maol Choluim* and means 'a follower of St Columba of Iona'. And so in Gaelic, *maol* became a term for a monk, 'a bald man', and so Melrose was 'the promontory of the monks'.

Cuthbert and his servant arrived at the monastery with a great deal more dignity than I had managed, but I believe he entered it at approximately the same place. Hidden in woodland to the east of an old track that was once a drove road in the eighteenth and early nineteenth centuries lies all that remains of the fabric of Old Melrose. Across the narrow neck of the peninsula, the monastic vallum was dug and its shallow ditches on either side of a central bank can still be made out amongst the lush green ferns and the willowherb. Much deeper when Cuthbert first saw it and almost certainly topped by a wooden palisade rammed into the compacted earth of the upcast, it marked a boundary between worlds. To the west lay the world the young man was leaving – the fields, farms and woods where men and women toiled, where warbands rode and killed, and where devils lurked in the dark places. Beyond the vallum was a place of light where the peace of God bathed the land, where the air was daily purified by the power of prayer and where holy men walked. Old Melrose was a portal, a place where monks strove through privation and contemplation to know the mind of God, where they looked with longing to the sky, hoping to approach Him and His angelic hosts more closely.

When I came at last to the vallum, I saw that it was crossed in two places and guessed that where I stood, the point at which the modern, tarmac road runs up to Old Melrose House, was the original gate, what Cuthbert would have understood as the first door to paradise. It lies closer to the track from Monksford. Here is Bede's account of what took place when Cuthbert and his servant arrived:

Now Boisil himself, who was standing at the gates of the monastery, saw him first; and foreseeing in spirit how great the man whom he saw was going to be in his manner of life, he uttered this one sentence to those standing by: 'Behold the servant of the Lord!' thereby imitating Him who, looking upon Nathaniel as he came towards Him, said: 'Behold an Israelite indeed in whom there is no guile'. Thus is wont to testify that pious and veteran servant and priest of God, Sigfrith, who was standing with others near Boisil himself when he said these words . . . Without saying more, Boisil forthwith kindly received Cuthbert on his arrival, and when the latter had explained the reason of his journey, namely that he preferred the monastery to the world, Boisil still more kindly kept him. For he was Prior of that same monastery. And after a few days, when Eata of blessed memory arrived, who was then a priest and the abbot of the monastery and afterwards both abbot and bishop of the church at Lindisfarne, Boisil told him about Cuthbert, declaring that his mind was well disposed, and obtained permission from him for Cuthbert to receive the tonsure and to join the fellowship of the brethren.

Much moved to find myself standing almost certainly in the same place where this momentous exchange took place, I sat down by the wrought-iron railings of a grass park to re-read Bede's account of the meeting at the gates. Aside from any other interpretation (was the entry of Cuthbert to Old Melrose agreed beforehand? Was there a payment?), what struck me most forcibly was the role, presence and voice of Boisil.

Probably a Gaelic-speaking Irish monk who came with Aidan from Iona when he founded the mother house at

Lindisfarne in 631, Boisil is sometimes credited with the establishment of the first community at Old Melrose. But as Bede was careful to note, he was not abbot and not able to admit Cuthbert without permission from Eata. Unlike his contemporaries, Boisil may have adopted his name. It is a Gaelicised version of Basil. An early bishop who led a famously ascetic life in the provinces of the Near East, what was becoming the Byzantine Empire in the fourth century, St Basil fought against heresy but eventually died, exhausted and enfeebled by the fasting and rigours of his exemplary life.

Boisil's own piety was much revered, and not only by Cuthbert. Three miles to the south of Old Melrose lie the villages of St Boswells and Newtown St Boswells, both named to honour the old monk. Their local pronunciation seems to get closer to him. Borderers call St Boswells, Bowzuls. At Benrig, a pretty hamlet half a mile from the older village, there once stood St Boisil's Chapel, only demolished in 1952. And a further link with Old Melrose was embedded in an ancient place-name. St Boswells used to be know as Lessudden, and it means 'the place of Aidan'. How and when one was supplanted by the other is long forgotten, but what the names signify was probably ownership. At some point in the history of Old Melrose, beginning soon after Cuthbert's coming, land was gradually gifted to the monks in return for certain privileges.

In the early Middle Ages and beyond, the idea of holy ground was not metaphorical. Where holy men had walked and prayed was literally physically sanctified and if a body was buried inside the monastic precinct of Old Melrose, and indeed many other sites where monks or priests had defeated demons and made the land sacred, it was believed that the earth itself would wash away mortal sin as the flesh rotted down to bone. Wealthy people were prepared to give lavish gifts in return for burial beyond the monastic vallum.

Much later, Anglo-Norman noblemen sometimes endowed a monastery on condition that they be admitted to the community as novices as the end of their days drew near. That meant they were guaranteed burial inside the sacred precinct, perhaps even under the floor of the church and close to the high altar. This was known as taking vows *ad succurrendum*, which translates as 'at the run' or 'in a hurry'. Robert Avenel, Lord of Eskdale and Richard de Morville of Lauder both became novices at the new abbey of Melrose and were buried under the floor of the church. Such privileges did not come cheap, and in 1216 Sir Alan Mortimer made over half of his estate to the old diseart that became the abbey of Inchcolm in the Firth of Forth if the monks would allow him to be buried in the church.

After re-reading Bede's account of Cuthbert's coming to the monastery, I walked through the sunshine up the road to Old Melrose House. It sits close to the highest point of the plateau and stands at what was the heart of the seventh-century community. In a copse beyond the house lie the ruins of an early medieval chapel dedicated to St Cuthbert. Under the shade of the trees, there is nothing much to see now; the only animation came from a group of small and excitable pigs (perhaps they thought I had come to feed them) who woke from their slumber in the warm afternoon to run around their pen as I passed.

Encircling the house and its knoll are wide, tree-fringed grass parks with nothing upstanding, no archaeological remains of any kind. It was as though the old monastery had never existed and the land had always been the leafy policies of a country house. Horses were grazing contentedly and nearby someone had set up showjumps in a practice area. But some sense of what once stood there can be found in Bede's work. He described Old Melrose's mother house of Lindisfarne as a very simple group of monastic cells probably made from wattle and daub with a wooden church near the centre of the precinct that had at first been roofed with reeds. There were probably one or two communal buildings, a cemetery and several fields for crops and pens for animals. All was enclosed by a monastic vallum.

The arrangement for Old Melrose is likely to have been similar: its wooden buildings all of wood; the monks' cells made from wattle and daub walls, and the chapel and other communal spaces from timber, all perishable. However, under the lush grass my old friend Walter Elliot has traced the watermark of sanctity.

Using an ancient method, Walter has brought the site back to life, discovering a great deal about what the monastery may have looked like and indeed what happened on the river peninsula long before the monks came to the promontory.

Walter is a diviner. Through his business as a fencing contractor, he found he was able to pinpoint all manner of underground features without the use of a shovel, using instead two metal coat hangers (or pieces of fence wire bent at right angles) held loosely in his fists. Much less energy-sapping and time-consuming than digging, and far more precise. Post-holes show up very readily, as well as water courses and pits, and as his fascination with history and archaeology developed, Walter began to walk across sites with his rods in his fists where he suspected there were the remains of buildings now lost in the grass. And he finds them. I have seen him do it time and again. And as they swing around, what the rods tell him is confirmed by subsequent investigation. Why it works, no one (including Walter) is sure. But it does.

Particularly in dry summers, aerial photographs can reveal what is invisible on the ground. In 1983, the Royal Commission for Ancient and Historic Monuments initiated a survey that included Old Melrose. Studying the subsequent photographs, Walter noticed a series of faint circular markings in the grass parks to the north-west of the wooded knoll, the site of St Cuthbert's Chapel and the estate house, that he reckoned to be the centre of the old monastery. When he walked slowly over the park with divining rods, quartering the ground carefully and noting and pegging what showed up as ground disturbance as he went, Walter found that the faint circles were in fact low banks and concentric rings of post-holes that enclosed areas that varied in diameter from approximately thirty feet to eighty feet. And beyond the edges of these circles were rows of rectangular pits that very closely resembled the outlines of graves.

Across Britain, archaeologists have discovered traces of similar structures. It seemed highly likely that on the site of the monastery Walter had found something much older, what is known as a wood henge. The most famous lies two miles to the south of Stonehenge in Wiltshire and it was first detected as faint circular markings by an aerial survey carried out in 1925. Later investigation revealed that Woodhenge was raised in the third millennium BC, and was still a focus of worship and ceremonial of some unknowable sort in 1800 BC.

Far easier and faster to build than stone henges, the function of these places was mysterious – but their essential nature unambiguous. Henges appear to have been first dug in Orkney as ditches and banks marked around by standing stones (more plentiful than trees in the Northern Isles) that served as temples of some kind. And just like the monastic vallum at Old Melrose and elsewhere, the circles of felled trees or quarried stones and their banks and ditches were set up as a means of separating the sacred ground from the temporal world around them. The simple duality of inside and outside was already present, something that persisted well into modern times. The mysteries of the mass took place behind a screen in abbeys, churches and cathedrals for many centuries and were heard but not seen by the lay congregation. Something of a similar sort may have taken place thousands of years before at henges. Those inside enacted rituals and sacrifices, probably chanted and played music, lit fires in the winter darkness and processed out through a throng of people who had heard but not witnessed the ceremonies.

If Walter Elliot has indeed discovered a series of wood henges at Old Melrose, that means something simple and very moving. The river peninsula had been revered as a holy place for more than two or three thousand years before the monks came from Lindisfarne. And if the rectangular pits arranged around them like the spokes of a wheel are indeed graves, then the desire to be buried in sacred ground was neither a purely Christian practice or new to the early Middle Ages. And what is even more intriguing are the reasons why a prehistoric, pagan culture wanted to bury its dead close to sacred ground.

There was no dubiety in the Christian era. Robert Avenel, Richard de Morville and many others believed absolutely that their internment inside the precinct at Melrose would eventually wash away their earthly sins and greatly enhance and ease their passage from this life to the next. And what happened after death obsessed the minds, practices and beliefs of the early monks and saints like Aidan, Boisil and Cuthbert. It seemed that all of their waking lives, and even their dreams and visions, were a preparation for the afterlife and the mortification of the flesh, the hours and days of prayer and vigil an attempt at cementing certainty, the sure and certain means of gaining entry to the Kingdom of Heaven. When on Brotherstone Hill Cuthbert saw a shimmer of angels bear Aidan's soul up through the clouds, it was nothing less than the justified reward for a saintly life of sacrifice led in the service of God and His truth. And what were the privations and sacrifices of a brief mortal life compared with the promise of an eternity in glory?

It may be that Walter Elliot has discovered hints of a remarkable continuity. Are the rings of graves, if they are graves, around the wooden henges at Old Melrose evidence of a pagan belief in an afterlife? They certainly demonstrate that our prehistoric ancestors thought that the place of burial (almost certainly for an elite, just as in the Middle Ages) was important and to be planted in holy ground was not an end but a transition. Otherwise, why bother? And it is significant that these henges were raised in the sight if not of God then certainly of a place of gods, Eildon Hill North.

As I wandered around the sunlit grassy parks where Walter found the circles of post-holes, I found myself musing on another transition, no less profound, if difficult to quantify. It is commonplace to assert that in the early twenty-first century we in Britain live in an increasingly secular age. Churches are everywhere closing and finding new uses, congregations are shrinking or merging, and it may be that for the first time in human history we are living in a society where most people do not believe in an afterlife, or at least postpone consideration of that until time begins to press on them. And more, does this not mean that we now fail to prepare for death in any clear or meaningful way?

For a long time, I found myself ducking this process with some woolly clichés, something along the lines of 'I'll just go on as always, only maybe a bit more slowly (daytime naps) but I will just keep on keeping on'. That is little better than a New Year resolution and not a preparation for anything. All of this is difficult territory and probably not worth mapping out with statistics and attitudes that are near impossible to measure. Instead, I began thinking much more seriously at Old Melrose about what my own feelings and attitudes are as I approach my eighth decade. Perhaps this is a second gift from Cuthbert.

As they often are, my thoughts were drawn back to the ghost in my heart, the shade of a little girl I was never allowed to know but who carried some of my blood and my genes. A few days ago, I noticed that the dry summer had turned the crown of Hannah's Tree brown. Through lack of rain from May to July, it had been struggling. We had to keep it alive, and Lindsay did, because the rowan tree was planted behind Adam and Kim's house in memory of their first child, our first granddaughter. Hannah was still-born, a clot having blocked her umbilicus and starved her to death. I find this difficult to write because the sadness floods back, the tears come, and nothing has hit any of us harder. Her funeral was the blackest, most difficult day of our lives and the image of her little wicker basket coffin will stay with me until the end of my days. It is all I have of her.

As I followed Cuthbert through a landscape so familiar to me and so much a part of my family's experience of life, I found myself thinking a great deal about the wee girl, about those who would have been part of her, those who made me and meant so much to me. Sometimes I can't seem to shake this bitter darkness. It is not her death I mourn so much as the complete loss of her future, and eventually a past. She had neither. When contemporaries die, we remember our shared past with them, and through the clouds of grief there is something to latch on to, something even to smile about. But we have nothing but loss with Hannah. She would have been three years old now.

There are photographs of Hannah Lindsay Karen Moffat, and Adam and Kim have a lock of her black hair, a handprint and other sad shreds of what she was and might have been, but I am so fragile about this that I have not yet been able to bring myself to look at these pictures and her things. Lindsay went to the hospital to see Hannah and told me she was perfect, but I feel the image of the little baby will haunt me. It is a variety of moral cowardice I am not proud of, but the memory of the profound sadness of two years and more ago (I cannot even bring myself to work out the precise date of her death; all I remember is that I had to go to London the day after she died to make a speech) is enough for me to cope with for now.

Little more than a year after Hannah died, her sister was born. One of the many joys of Grace is a prosaic but important one: she inherited all of the clothes and some of the things that were bought for her older sister. It took great courage from Kim and Adam to try for another baby so soon after Hannah's death, but it was the right thing to do because forcing themselves to look to the future, and the happy fact that Grace is a girl, made the pain a little easier to bear. A surprising and even more painful part of what happened does stay permanently with me, something more that was lost. Hannah was dark-haired like Kim, and Grace is very fair, like Adam, and I would so liked to have seen them side by side, Hannah teaching her little sister about the world. We shall never forget her, but we have lost her – almost completely.

Before we sat down at the table on Christmas Day last year, Adam asked if we had any tea-lights and his mum found some in the boiler room. He then lit and placed one in a small metal holder with Hannah's name engraved on it. He and Kim are determined that she will not be forgotten or her presence left out of important occasions and I had to swallow hard at the other end of the table as her little sister shrieked with excitement. Sadness amidst joy is simply how life is, I suspect, and I was glad to remember the wee lass and what might have been. As the years pass, the pain of her death grows no less sharp.

Instead of putting Hannah's death and the cruel robbery of her future to the back of my mind, I wanted to honour her by learning something from it. She is dead and I cannot reach across the snows of eternity and help her, but perhaps the wee lass can help me. I began to think more and more that we should keep the dead close to us. Hannah is buried on our farm in the garden of her parents' house, the place we all made and invested with memories, and her first lesson for me is the firm decision that I will be buried alongside her, where she is waiting for me.

When we all decided that Grace should have a naming party (her parents are not believers in any religion), I asked to say something. What follows was an early attempt I made to describe a different sort of afterlife, and unusually I wrote it down, wanting to be precise and clear. It did not work out that way.

Katherine, the lovely humanist celebrant who had married Kim and Adam, had also come down to the crematorium to help us with Hannah's funeral and she seemed the right person to be with us on Grace's day. Unexpectedly, she began by saying something about Hannah and suddenly, waiting to speak, I was overwhelmed by a tide of sadness at a moment when I wanted to try to bring the living and the dead together. Having managed to hold back the tears until I reached the last paragraph of what I had to say, I simply had to stop for a long moment to gather myself. Afterwards, I was disappointed, since I felt that what I had to say might be the beginnings of learning something from Hannah's death, something not about immortality but about a life after death, and these attempts were diluted when grief engulfed me. Here is the text of the short speech I almost failed to make:

As Grace Moffat begins her journey into the future, she takes the past with her. It is an immense past, reaching far back beyond memory, much of it lived amongst these sheltering hills and rolling river valleys. Her people walked their lives under the big skies of the Borders, skies that seem to fascinate her now. Grace's gaze is often upwards, beyond the immediate, her deep blue eyes lifted towards to the blue yonder of the sky. The ghosts of Grace's past will walk beside her as she skips down the track to the stables or walks out with the dogs on a summer morning.

This is Grace's day, a day of names, the names of memory and the names of love. Grace Moffat is the great-great-grand-daughter of Bina Moffat. In the late nineteenth century she too was born on a Borders farm, at Cliftonhill near Kelso, not far from here. Grace will come to know what Bina knew – the snell winter winds whipping off the hills, the butter-coloured sun spreading over the dawn fields, the summer bummies buzzing in the warm breeze, the robins picking at the midden in the hungry months of the late winter. Although Bina was born in 1890, 126 years ago, Grace will come to know something of her life.

Ellen Moffat is her great-grandmother. Hawick-born and raised as one of seven sisters and a solitary brother, she lived in the body-warmth of a busy industrial town, working in the rattle and clack of the textile mills and in the culture of the close community of the common riding, where women could put men in their place with a single freeze-frame, old-fashioned look. Her great-granddaughter has yet to perfect that necessary art, but like Ellen Moffat, Grace Moffat has a smile that could light a darkened room.

On this day of names, Grace's grandmothers, Lindsay and Karen, have wrapped their girl in love and memory. Grace wears Lindsay's own christening gown and she is wrapped in a shawl made by Karen's mother, Kim's much loved Nana. The grannies have much more to give her, but the detail of that will be far beyond the understanding of any man. Love has secrets, and Grace and her grannies will giggle and smile while father and grandfather shake their heads and shrug.

Kim Moffat gave Grace Moffat the greatest gift. Her bond with her baby is unbreakable, forged in the gazes and glances of love, in the fitful sleep of broken nights, in the burble and chatter of wakefulness. Grace is beautiful and Kim made her beautiful.

Grace has changed us all, directing us to a different future, but she has changed Adam utterly. I have never seen a father so involved, so sharing of the necessary burdens, so confident with his wee one in his protecting arms, so attentive and so expert. You are a fantastic father, Adam, and your mum and I are so proud of you.

But there is one obligation we can share as a dad and grandad. We should help Grace remember things she cannot know. You should tell her about the generations that have gone, about her great-grandparents, Jack and Ellen Moffat, Malcolm and Helen Thomas, while Kim can tell her about her own beloved Nana. And I'll make sure that Bina Moffat takes Grace Moffat's hand as these two children of the land walk around the edge of a circle that leads from Cliftonhill *in 1890 to this farm in 2016. And they will share a secret: both will know that on moonless nights, you can find your way by starlight.*

Keeping the dead close can make the future come alive. No matter how painful it is to remember the death of Hannah – and I have struggled all morning with tears as I wrote this – we make our dead live much longer if we keep them alive in memory, anecdote and even with the passing on of inherited characteristics, names, hair colour, facial expressions, phrases and much else. In an increasingly atomised society, shaken up and slackened by social mobility, these precious threads that bind the present to the past are often broken. Too few, for example, can name all four of their grandparents. Some years ago, I was involved in an ancestral DNA project where we asked a large sample exactly that question and I was shocked at how few could answer without first doing some research.

In order to keep the dead closer, we do not need to rewind history, return to an impossible bucolic Brigadoon of farming folk who never went anywhere. We just need to remember more and better, and change the way we think. Some years ago I did an immersion course in Scots Gaelic and was struck by how well and simply its usage caught the body warmth of the communities in the Western Isles where the old language is still heard. When Gaelic speakers meet someone new, they don't ask where they live or what they do, the sort of standard exchange between strangers at parties or functions. Instead, they enquire, '*Cò as a tha thu?*' Which loosely translates as, 'Who are your people?' That is surely a much better question, one that will elicit a personal, perhaps even unique response. There are many teachers, road sweepers and software designers in the world, but in telling someone about your family and not your job or your address, you fill in a background that is yours alone. And you keep the dead close.

Death loomed much closer for Cuthbert and the monks at Old Melrose. In an age before all the blessings of modern medicine, life in the seventh century was considerably shorter and could end abruptly and painfully. A great pestilence several times stalked the land. In 541 what is known as the Plague of Justinian broke out in Constantinople. A version of bubonic plague, it devastated populations and its first wave killed about twenty-five million people, perhaps 13 per cent of the world's population at that time. Historians believe that its spread to Britain fatally weakened post-Roman society and made it easier for the Angle and Saxon invaders to defeat native kings and take over their territory.

Plague broke out again in the middle of the seventh century and the pandemic is thought to have affected Britain particularly severely in the decades after 660, not long after Cuthbert dismounted at the gates of Old Melrose. Bede wrote simply 'the pestilence came', and to some its visitations seemed apocalyptic, a manifestation of God's anger at the sinfulness of men and a prelude to the end of days, the Last Judgement.

Not only did the plague encourage the devout in their prayer and privation, making their preparations for a life after death more urgent, it also encouraged them to think about what the world would be like after the apocalypse, what was sometimes thought of as the Second Creation. Being closer to God than ordinary mortals, saints were believed to have insights, to understand something of how the world and all its creatures would be remade anew by the hand of the Almighty and how heaven on Earth would be restored.

Cuthbert loved animals and several of his miracle stories tell how they helped and looked after the saint. Both Bede and the Anonymous *Life* recount his journey from Old Melrose to the cliffs of the Berwickshire coast to visit the Abbess Aebbe, the sister of King Oswiu of Northumbria, but the latter's is more eloquent and atmospheric:

He came to the monastery which is called Coldingham, in response to the invitation, and remaining there some days, did not relax his habitual way of life but began to walk about by night on the seashore, keeping up his custom of singing as he kept vigil. When a certain cleric of the community found this out, he began to follow him from a distance to test him, wishing to know what he did with himself at night. But that man of God, approaching the sea with mind made resolute, went into the waves up to his loincloth; and once he was soaked as far as his armpits by the tumultuous and stormy sea. Then coming up out of the sea, he prayed, bending his knees on the sandy part of the shore, and immediately there followed in his footsteps two little sea animals, humbly prostrating themselves on the earth; and, licking his feet, they rolled upon them, wiping them with their skins and warming them with their breath. After this service and ministry had been fulfilled and his blessing had been received, they departed to their haunts in the waves of the sea. But the man of God, returning home at cockcrow, came to the church of God to join in public prayer with the brethren.

The above-mentioned cleric of the community lay hidden amid the rocks, frightened and trembling at the sight and, being in anguish all night long, he came nigh to death. The next day he prostrated himself at the feet of the man of God and, in a tearful voice, prayed for his pardon and indulgence. The man of God answered him with prophetic words: 'My brother, what is the matter with you? Have you approached nearer me, to test me, than you should have done? Nevertheless, since you admit it, you shall receive pardon on one condition; that you vow never to tell the story so long as I am alive.' The brother made the vow and kept it afterwards and departed with his blessing, healed. But after Cuthbert's death, he told many brethren how the animals ministered to the saint, just as we read in the Old Testament that the lions ministered to Daniel.

What terrified the young monk who spied on Cuthbert was that he had seen something he should not, a glimpse of what might be the Kingdom of God. The behaviour of the sea otters was a memory of the harmonies of the First Creation, when God walked in the garden. After the serpent hissed to Eve that she should eat the forbidden fruit, and she and Adam realised they were naked and were ashamed, much was lost and prophets longed for a time when once more the 'lion would lie down with the lamb'.

It was the power of Cuthbert's prayer and privation, his near-immersion at dead of night in the ice-cold waters of the North Sea, that brought forth the otters from the waves. And that was the fundamental lesson of the miraculous story. Night vigil, psalm singing, fasting, prayer, the pain of extreme cold and constancy of faith would eventually right the balance of Creation and compensate for the original sin committed by Eve and Adam. The old saints and their fellows underwent all sorts of painful tests not only for the sake of their own salvation but also to implore God to restore the lost harmonies of Eden for all mankind.

In both Bede's and the Anonymous *Life* there are several stories of animals helping (and initially hindering) Cuthbert. As he and a companion travelled up the River Teviot, perhaps by boat, and into the hills on a mission of conversion, an osprey or perhaps an eagle brought them 'a large fish', almost certainly a salmon. When the boy took it for themselves, the man of God rebuked him and instructed that the fish be divided and half given to the eagle: 'Why did you not give our fisherman a part of it to eat since he was fasting?' On modern Lindisfarne, Cuthbert's bond with animals and birds in particular is fondly remembered. Perhaps because he insisted that its nests on Inner Farne be untouched, the eider duck is known as Cuddy's duck.

When Cuthbert first retreated to make a hermitage on Inner Farne, he dug and trenched such land as there was so that he could grow his own grain and vegetables and so reduce the frequency of visits from the outside world. Two ravens arrived and began to tear at the thatch of a shelter built for visitors in order to build a nest. It must have been springtime. 'With a slight motion of the hand', Cuthbert shooed them off and told them to cease. When the ravens ignored him, the saint invoked the name of Jesus Christ and banished them from the island. And then this fascinating exchange took place:

> . . . after three days, one of the two returned to the feet of the man of God as he was digging the ground, and settling above the furrow with outspread wings and drooping head, began to croak loudly, with humble cries asking his pardon and indulgence. And the servant of Christ, recognising their penitence, gave them pardon

and permission to return. And those ravens at the same time having won peace, both returned to the island with a little gift. For each held in its beak about half a piece of pig's lard which it placed before his feet. He pardoned their sin and they remain there until today. Most trustworthy witnesses who visited him, and for the space of a whole year greased their boots with the lard, told me of these things, glorifying God.

On the windswept rock that is Inner Farne, Cuthbert's closeness to God, his sanctity and patience, had created a tiny Edenic enclave where the harmonies of the Second Creation were prefigured. These were the beginnings of a tradition that found its most famous flowering in the life of St Francis of Assisi. While it is easy to dismiss these tales as early medieval versions of Dr Doolittle, they suggest a different, perhaps simpler and more profound way of thinking about the world, the whole world and not just the way men and women have exploited and bent it out of its natural shape.

The remarkable summer of 2018 ignited a more intense debate about global warming and what our impact has been on the climate. These issues are vast, many of them technical, and the politics that reverberate around them often just as extreme as the changing weather. Instead of a Second Creation, we or our grandchildren may witness an apocalypse. And where shall we run to then?

Attitudes clearly have to change and soften, but I fear it may take too long for the extremes of the climate to have that effect. Instead, the stories of Cuthbert, other early saints and their animals around them may offer a more immediate way of understanding how the world might work better. Over the last twenty-five years, my own attitudes have changed radically because of my family's direct interaction with animals, our efforts to understand them and to help them understand us.

When we first came to our farmhouse, I suggested that our young daughters might like to ride. Close at hand there was a good riding school and, if over the space of a year they enjoyed lessons there and came to understand the hard work and responsibility of looking after horses, then maybe we would find them their own ponies. This was not the reaction of well-heeled city dwellers converting to country ways. I had grown up in the Borders, with its strong equestrian traditions borne out of the annual festivals held in each town that are known as common ridings. One way or another, most people, regardless of status or income, who wanted to ride had their own horse and were usually able to find grazing and stabling. Equestrianism in the Borders was not and is not the preserve of the posh or entitled; it is woven into the fabric of society and has been for centuries.

And so in 1995 there began a long family love affair with horses that has reached new and different heights. My daughters gained much by riding; both have a stillness, sympathy and gentleness that comes from close relationships with these noble creatures. And the usual self-centred obsessions of teenagers were much tempered by the love and responsibility they felt for their horses. After they went to university and gradually gave up riding, my wife began to breed horses, a tremendously demanding and risky undertaking. Now she is the owner of a very beautiful and enormously talented dressage performer. This extraordinary little coloured mare is a double Scottish and British champion at novice level, and barring injury, to say nothing of tempting fate, we believe that she may have a long and illustrious career ahead of her. But the point here is not to document success (there has been plenty of failure, too) but to make a simple connection with Cuthbert's relationship with the sea otters and the birds. Equestrianism is the only sport where women and men form a similar relationship with an animal, where they need to communicate with the horse in complex ways, understand its needs in the absence of speech and develop powerful bonds of trust, even love.

Horses change people. They have changed me. And observing and coping with their behaviour can be a means of better understanding more of the natural world, as well as a way of softening attitudes. No one who genuinely likes horses is anything other than thoughtful around them (of course, there are some whose sharp competitive instincts are not attractive, but even they usually know enough not to communicate ill humour if they want a horse to perform for them) and in grooming, management and riding they try to avoid sudden movement, loud noises or anything upsetting. It is well understood that people with mental health issues can find some peace helping in a stable yard, and in a remarkable development three American airlines now allow miniature 'service' horses to travel with passengers who need the reassurance of their presence. Being close to these gentle animals is a good beginning in life and also a harking back to an older Britain – long before the racket and rush of cities, and the industrial revolution – a time when our land was quiet and green, and completely dependent on horses for traction and transport. A time when we had to understand and communicate with these animals.

Soon after we moved to the farmhouse, we built three loose boxes by the burn at the bottom of the slope. They became home to my daughters' horses and a dapple-grey Connemara mare my wife rode. Gradually our small farm became a hub of animal activity of all sorts. The horses' muck heap brought flies that in turn attracted swallows, who began to nest in the loose boxes. Around the paddocks close to the farmhouse we planted hedges and trees that became home to more birds, hedgehogs and rabbits, and sometimes in the late evenings we saw the grey shapes of roe deer grazing the farthest pasture. In the dry summer of 2018 many house martins came, their white striped tails flashing in the sun, and in the warm evenings we watched enthralled at the daredevil aerobatics of thirty or forty birds swirling around the sky above the steading and the stables, feeding on flies, feeding their chicks under the eaves of the house, feeling the elemental joy of being alive.

I am well aware that being in the midst of all this vibrancy and joy is a privilege, something not readily accessible to many. But it is a choice. All of our dwindling resources are spent on keeping this farm and its animals going, and it can be a very expensive business. We take no holidays because we cannot afford to and, in any case, who would look after our fifteen horses, three dogs and maintain the fields, fences, tracks and all the unexpected things that managing a farm entails? Of equal concern is our physical ability to continue to look after the animals without resorting to shortcuts. Nearing the end of our seventh decades, we have to limit the amount of time we spend in the evening complaining about aches, small breaks (toes, the occasional rib) and near-constant arthritic pain because if we did not, we would talk of little else.

Cities and large towns have largely separated Britain's people from the animals they used to live close to and now all most people see are pets and occasionally rats, mice and urban foxes. It is certainly a great loss, a wide gap in experiencing the wholeness of the world, but what I had not understood was the spiritual nature of our daily contact with our horses, dogs and the wild animals around us. Cuthbert's stories and the notion of the harmony of the Second Creation reminded me that even non-Christians can understand the value of being much closer to the creatures with whom we share the planet – and who make it habitable.

By mid-afternoon, the sun had climbed high above Old Melrose and, having forgotten to bring a hat, I sought the shade of the woods to the north-west of the river peninsula. Magnificent trees towered above me: wellingtonias, Scots pines, oaks, spear-straight poplars, shimmering aspens and others I could not identify, all of them the glorious consequence of the selfless foresight of late nineteenth-century planting. Through the woods, I could see the glint of the Tweed as it began to turn, its course first forced east and then south by the glowering red sandstone mass of the river cliff. The great river has its own watery map, the names of pools, beats and rapids usually conferred by fishermen. Where the Tweed turns most sharply, the uncountable millennia of spates and battering ice floes will have scoured its bed and made it very deep, and even in that rainless summer the pool known as the Crom Weil looked dark and dangerous, as unseen currents churned below the placid surface. *Crom* is from Old Welsh, the language of the Tweed Valley before Cuthbert's people brought Early English, and it simply means 'a bend'. *Weil* is cognate to 'wheel' and may refer to the motion of the currents. Or it could be related to the word *wael* for 'a pool'. River names are often the oldest in the landscape and their origins difficult to parse.

To the south of the plateau of Old Melrose, the pools remember the centuries of the monks. Where there is a sliver of a long, low river island, the map plots Halliwell Stream, and as the river turns slowly towards Monksford, the pool is marked as Holy Weil. These may have been the shallower reaches of the Tweed where mortification of the flesh was practised.

Immersion in bone-chilling water was common and the most extreme example is the story of Drythelm, a monk at Old Melrose in the decades on either side of AD 700. As a layman he had fallen into what was probably a coma and, according to Bede, was resurrected as though from death. To astonished audiences, he then related a vision, a journey in the company of 'a handsome man in a shining robe . . . to a very broad and deep valley of infinite length', what sounds like an early visit to a version of purgatory. Because many believed that Drythelm had glimpsed what lies beyond death, people often came to Old Melrose to listen to his account of his vision, including King Aldfrith of Northumbria. At the close of his passage about Drythelm, Bede described how the monk often immersed himself in the River Tweed, no matter the season:

This man was given a more secluded dwelling in the monastery, so that he could devote himself more freely to the service of his Maker in unbroken prayer. And since this place stands on the bank of a river, he often used to enter it for severe bodily penance, and plunge repeatedly beneath the water while he recited psalms and prayers for as long as he could endure it, standing motionless with the water up to his loins and sometimes to his neck. When he returned to shore, he never removed his dripping, chilly garments, but let them warm and dry on his body. And in winter, when the half-broken cakes of ice were swirling around him, which he had broken to make a place to stand and dip himself in the water, those who saw him used to say: 'Brother Drythelm (for that was his name), it is wonderful how you can manage to bear such bitter cold.' To which he, being a man of simple disposition and self-restraint, would reply simply: 'I have known it colder.' And when they said: 'It is extraordinary that you are willing to practise such severe discipline,' he used to answer: 'I have seen greater suffering.' So until the day of his summons from this life he tamed his aged body by daily fasting, inspired by an insatiable longing for the blessings of heaven, and by his words and life he helped many people to salvation.

Now, this borders on Pythonesque pastiche, but nevertheless it has the whiff of authenticity. Immersion carries overtones of cleansing and repeated re-dedication by baptism, and as the water reached his loincloth it will also have been very effective in suppressing the desires of the flesh, the early medieval equivalent of a cold shower. While Drythelm shivered amongst the ice floes, more gentle sorts of devotion also took place on the river peninsula.

As Bede remarked, Cuthbert could have chosen to enter the monastery at Lindisfarne but instead chose to come to Old Melrose explicitly because of the reputation of Boisil. By the time he had retreated for his first period at the hermitage on Inner Farne, the saint remembered the old man very fondly when he spoke to visitors:

I have known many of those who, both in purity of heart and in loftiness of prophetic grace, far exceed me in my weakness. Among these is the venerable servant of Christ, Boisil, a man to be named with all honour, who formerly in his old age, when I was but a youth, brought me up in the monastery of Melrose, and, amid his instructions, predicted with prophetic truth all the things which were to happen to me.

Cuthbert and Boisil became soul-friends. A translation from Irish Gaelic, *anam-cara*, it is an attractive description of close affection and kindness, and it carried a further sense of one being a teacher and confessor, the other a pupil and disciple. As time went on at Old Melrose, and Cuthbert's faith and knowledge accumulated, there seems to have grown a strong bond, as each monk held up a mirror to the other's soul. But in 664, when Cuthbert had probably been at Old Melrose for thirteen years, the soul-friends were sorely tested.

In that year Adomnan, Abbot of Iona, recorded the coming of a great pestilence, what was known as the Yellow Plague, a recurrence of the devastating Plague of Justinian. Cuthbert was infected and Bede wrote of 'the swelling which appeared in his thigh', a symptom of bubonic plague. But the young monk survived, as some did, saying, 'And why do I lie here? For doubtless God has not despised the prayers of so many good men. Give me my staff and shoes.' For the rest of his life, Cuthbert felt the painful effects of this terrible illness, but he was more fortunate than his soul-friend.

Much older, Boisil knew that when he too was infected, he would die quickly, having seen others consumed by plague, and he predicted that the end would come in seven days. But instead of seeking consolation or sympathy, he told Cuthbert that he should lose 'no opportunity of learning from me so long as I am able to teach you'. The young monk asked Boisil what would be best for him to read for the days that remained, and they agreed on the Gospel of St John. 'I have a book of seven gatherings of which we can get through one every day, with the Lord's help, reading it and discussing it between ourselves.' Instead of a deeper analysis of the text, the soul-friends decided to speak only of simple things such as the 'faith which worketh by love'. And once they had completed their readings and discussions, Boisil died and 'entered into the joy of perpetual light'.

The reality is likely to have been very different, very grim indeed. Scientists and historians now believe that the Yellow Plague of the seventh century resembled the Black Death of the middle of the fourteenth century. The Italian writer Giovanni Boccaccio left a description of what happened when an individual was infected with that later plague: 'At the beginning of the malady, certain swellings either on the groin or under the armpits . . . waxed to the bigness of a common apple, others to the size of an egg . . . and these are the plague boils.'

These boils or buboes oozed blood and pus and the flesh began to blacken around them. As the disease raged through a victim's body, they suffered terribly with fever, sleeplessness, vomiting, diarrhoea and severe pain before the mercy of death took them. Cuthbert's soul-friend will have suffered these hideous symptoms before he reached his last day, which was spent 'in great gladness', as Boisil longed for the comfort of death, consoled as much as was possible by readings from scripture.

We will all face the moment of death alone, but instead of staring at a chill and bleak horizon of nothingness, Boisil's hand was held fast by his *anam-cara*, who was with him until his last moment, reassuring him as his eyes closed that a benign God waited to welcome him into eternity.

Unlike Cuthbert and Boisil, and in common with many others, I do not have the comfort of a belief in God or in heaven, but I do think that the prospect of holding the hand of a soul-friend as the breath finally slips out of me is a warming one. No one knows me better or loves me more than my wife. I love her too, and we have been together for more than forty-five years. Of course there have been good and bad times, but none of that will matter when I find myself in the imminent presence of the wide solitude of death because she will be holding my hand. I can go then.

My friend Richard Holloway has recently written wonderfully and movingly about facing death in his *Waiting for the Last Bus*. He makes the point that unlike the seventh century, when death from all manner of causes, including bubonic plague, stared Cuthbert, Boisil and their contemporaries in the face, we now live in a society that seems to deny its inevitability and does everything medically possible to preserve life. And while Richard would never presume to offer a prescription on how to deal with our deaths, he does make an excellent suggestion, something he himself has done. Since there is nothing one can do to avoid extinction, why not plan your own funeral? His view is that it should provide some comfort to know what will happen when you are buried or cremated, even if you cannot be there. Even though it feels strange to contemplate this, I think he is right. Boisil planned his last days, amidst all that gathering agony, and that seemed to give him – and his soul-friend – some consolation.

No doubt the old monk was also comforted by forgiveness. If Cuthbert assumed the role of confessor, and as his *anam-cara* he surely must have, then he will have given Boisil absolution for his sins. At the moment of death, I can understand why forgiveness is important. Even though I have no belief in God and do not therefore seek the forgiveness of anyone other than my family or close friends, I do not want to die unforgiven for all the mistakes and omissions I have made, some of them mean and selfish. As it draws to a close,

I want to feel that I have lived a decent life and done more good than harm.

Some time in 664, Cuthbert succeeded Boisil as Prior of Old Melrose, but the Anonymous *Life* and Bede's are unclear about how long he served in that office. '*Per aliquot annos*' is the phrase used, and it can mean 'for some years', either many or a few. It seemed that amongst his most urgent duties, the new prior walked the valleys and villages of the Tweed basin preaching the word of God on missions of re-conversion. The devastation of the Yellow Plague had caused many to return to the old pagan gods.

After Cuthbert's elevation in 664, the Anonymous *Life* contains a brief and enigmatic passage, 'but finally he fled from worldly glory and sailed away privately and secretly'. Since leaving the monastery without the permission of the abbot is a serious transgression that does not fit into the picture of an exemplary life, Bede glosses over this episode. But it seems very important: a moment in the narrative that jars and one that caught my attention. Why did Cuthbert leave Old Melrose? Did he have a breakdown after the death of his soul-friend? Or did he pine for the solitary purity of life as a hermit? One aspect is at least clear: Cuthbert must have left his monastery secretly by boat.

5

In the Arms of Angels

In the sixth, seventh and eighth centuries Irish monks, mystics and missionaries put their trust in God and sailed the wild seas off the western coasts in search of promised lands. Place-names trace their epic voyages. In the Faroese language, *papars* specifically means 'Irish monks', and the islands of the west and north remember their wanderings: Pabbay and Bayble in the Hebrides, Papa Westray in Orkney, Papa Stour in Shetland, Papey off Iceland, and many others. Faith took the saints great distances and the most famous and farthest travelled was St Brendan the Navigator. With fourteen companions, he sailed the North Atlantic through ice floes and past monsters and apparitions, before setting a course by Hesperus, the Evening Star, and making landfall on a shore cold with seals and angels.

Brendan's account of what he called 'The Journey to the Promised Land' strongly suggests that he found America 1,000 years before Columbus. In 1976 Tim Severin set sail from southern Ireland in the *St Brendan* to follow the course implied in the sixth-century account and, mostly relying on the same navigational aids as the Irish abbot, he reached the Labrador coast. The much shorter voyage of Columba from the north of Ireland to Scotland's Atlantic shore in the mid-sixth century recalls the sort of boat used by Brendan and Severin. Tradition holds that the saint made landfall on the western side of the tiny island of Iona in 563 at a bay known as Port na Curraigh.

Curraghs were made from ox-hides sewn together, stretched over a frame of whippy green rods (hazel was much favoured) to keep the hull taut, and caulked with wool grease. Rowing benches were wedged in as thwarts to make the canoe-like shape as rigid as possible. Tim Severin was surprised to find that his large sea-going curragh of the sort sailed by Brendan and Columba could scud across the waves at speeds of twelve knots, faster than some modern yachts. This happened because these simple leather boats were very light, had no keel to create resistance and consequently took a shallower draught, with smaller curraghs needing less than a foot of water to float them. Tossed on the big seas of the North Atlantic, Severin often felt a lack of something Brendan had in abundance – faith in God.

Construction was quick and easy once the ox-hides had been cured and repairs and patching were not difficult. St Brendan and his monks took spare ox-hides and wood with them on their voyage to the promised land. Many years ago I found myself in Cork and I went to a small, even ramshackle, boatyard on the banks of the River Lee, opposite the Beamish brewery, to watch Padraic O'Duinnin build a small curragh. Working alone, and never once using a measure of any sort, he built the boat in a morning, using treated canvas instead of ox-hide. When he had finished, he showed complete confidence in his skills by immediately floating the little curragh on the river.

Since Old Melrose and Lindisfarne were founded by Irish monks from Iona, there can be little doubt that they brought the simple technology of curragh and coracle (a smaller, round version used on lochs and smaller rivers) building to the banks of the Tweed. And equally there can also be little doubt that the great river was an ancient and medieval highway. Excavations at Trimontium, the great Roman army depot only a mile or so upstream from the monastery, have uncovered a steering paddle, probably intended for use on a raft, and also much evidence of the sort of bulk transport only possible on water, large wine amphorae and the like. In the twenty-first century, the only craft seen on the Tweed are used for leisure pursuits: the grey rowing boats of fishermen and the kayaks and canoes of white-water competitions. But in the past it was much faster and safer to travel and move goods on water, if at all possible. When Cuthbert stole away from Old Melrose 'privately and secretly', the Latin verb used by the writer of the Anonymous *Life* was '*enavigavit*'. He sailed away, almost certainly in a small curragh, allowing the currents of the Tweed to bear him from worldly glory, beating the dark water with his oars.

Cuthbert may have slipped out of the monastery '*occulte*', or secretly, in the dead of night, but I wonder if he had any notion of where he was going. In common with St Brendan, other Irish monks sometimes took to the sea in their curraghs as a means of surrendering themselves to the will of God. Wherever He willed them to be carried by the winds, the tides or the currents was where they would go. '*Fugiens*', Cuthbert was fleeing from the temptations and cares of the world, and I suspect that at first all he wanted was distance from those and the beginnings of a life of solitude. Curraghs are so light that a small one could easily be picked up by one man and the shallows at Monksford will have presented no difficulty. In less than an hour, the fugitive could have travelled a long way, passing the site of Modan's diseart at Dryburgh and rounding another loop of the meandering Tweed where St Boisil's Chapel would later be built.

The Anonymous *Life* deals with the flight of Cuthbert in one sentence, while in his prose *Life*, Bede ignores it. However, there exists another early source that might supply answers to the mystery of what happened to the man of God. Before he composed his prose *Life*, some time around 705, Bede wrote a shorter version in poetic metre, what is known as the Metrical *Life*, and in it there exists a passage that does not reappear in his later work. Cuthbert wished to avoid the praise of men and: 'prefers to wander over the secret tracts of a solitary place, where, with God as witness, he may be free, guarded from the fame of human praise'.

Between them, these two sentences allow a partial sense of what Cuthbert sought and also what feels like a precipitate departure from Old Melrose, but they do not explain why he gave up the office of prior to wander alone amongst the valleys, woods, moors and hills of the borderland. Perhaps politics and the events of 664 might offer a context, if not a direct motivation.

In that year, King Oswiu of Northumbria summoned his bishops, abbots and leading churchmen and women to Whitby on the North Yorkshire coast to resolve a dispute. The Church of Rome had devised a formula to fix the date of Easter, the most important festival in the Christian calendar. It was famously movable for reasons long forgotten: in the early centuries after Christ's death, the Church wanted to avoid a clash with the Jewish Passover, the time when Jesus was arrested, tried and crucified. Probably for reasons of remoteness, the Celtic Church had fallen into the habit of setting a different date, one that ignored the Passover.

King Oswiu had married Eanfled, a Kentish princess who kept the Roman dates and at Easter he found himself feasting alone. According to his wife, it was still Palm Sunday. But there were difficulties more fundamental than domestic disharmony. In the seventh century, and for a long time afterwards, Christians believed that Easter, their pivotal, defining festival, was the occasion of a great battle between God and Satan. And for God to triumph in that battle, all believers, all of his army of supporters, had to pray for victory – at the same time. It was a matter of sheer numbers.

There were other issues at stake. The Roman tonsure cut in imitation of the crown of thorns was preferred to the Celtic tonsure cut across the forehead from ear to ear with the hair grown long at the back, a probable harking back to the Druids. When he was first admitted to Old Melrose in 651, Cuthbert will have been tonsured in this way. Rome will have sniffed at such whiffs of the pagan past. In addition, Easter's date shaped the whole Christian year. All of the other festivals, such as Lent, Ascension Day and Pentecost followed from it and therefore the dispute was not only over one celebration: different dates meant two entirely different liturgical years.

The politics of what became known as the Synod of Whitby were brutal and when Oswiu ruled in favour of the Roman method of dating Easter, the monks of the Celtic Church simply departed. For Cuthbert, this decision and its timing tugged at the foundations of his faith. He had been instructed in the ways of Aidan and Boisil, and the asceticism of the Irish monks was burned deep into his beliefs. When his soul-friend was taken by the Yellow Plague in 664, the same year as the synod, Cuthbert's world was turned upside down. Not only would the young monk have struggled to come to terms with the loss of his teacher and confessor, he also assumed the weighty responsibility of the office of prior. In that role, he would have had to make sure that the changes insisted on at Whitby were enforced. Perhaps the older monks at Old Melrose resisted; perhaps Cuthbert tired of ecclesiastical politics and pined once more for the purity of the hermetic life where God governed all.

While I have never relished the idea of solitude, I well understand some of the instincts that prompted Cuthbert's departure from the hierarchy of the church, from the establishment that ran seventh-century Northumbria. For different reasons, I have always felt myself to be an outsider. When I left the body warmth of my upbringing in Kelso to go to university in those far-off days when working-class children with some ability had their tuition fees paid by the local authority and a maintenance grant from the Scottish Education Department, I felt I began to inhabit a no man's land between two worlds. It is a place I have never left.

When I found myself at St Andrews University, my speech switched from the Scots of hearth and home to a version of tidied-up English that belonged to what I sensed would be the future. Whatever that might be, it would happen in a language that never quite fitted my mouth, was not mine, and life would be measured and organised according to dictums and customs that were often foreign to me. But no matter its awkwardnesses – I suspect my early attempts at tidied-up English were occasionally hilarious, and I still make mistakes in pronunciation – social mobility was what my parents wanted for me, to have a better, more prosperous and interesting life than they had, and in my dad's phrase, to get a job where you don't have to take your jacket off. Both my mother and father were articulate, well-read and highly intelligent people but the opportunities that opened for me were not available to them in the 1930s. And so I felt I could not take any other course than to graduate and find a well-jacketed job, probably as a teacher (no sense in reaching too high) and climb up a few more rungs than they did.

Of course, I always came home. My bond with my family, especially my mother and my sisters, was strong. And while my dad and I fought, sometimes literally when I grew up and became bigger than him, I loved him. When we were not talking, we could always talk about rugby, and the council estate where we lived was very close to Poynder Park, Kelso's ground. It turned out that I had a gift for the game, perhaps the only natural talent I ever had. But when I went to university and came home to play for Kelso, I experienced the first real rupture with my past. Because I was a student, paid for by the taxpayers, and almost all of my team-mates were not, there was resentment. Maybe they thought I looked down on them. And because, to be frank, I was more talented, that resentment deepened, particularly at the beginning and end of the season. Those were the times when prestigious and well-attended seven-a-side tournaments took place in the Borders and selection for the Kelso seven was thought to be a privilege, a reward for slogging through the mud and rain of the winter. I was fast, big, could tackle and had good hands, as well as the ability to kick goals, so I was selected ahead of other stalwarts in the team and their annoyance sometimes erupted into foul play and fighting at training sessions. And at the sevens tournaments themselves, one or two of my team-mates would not pass me the ball, even though I was running free into space. It all became so sour that I eventually gave up playing when I was twenty-two, something I bitterly regret. You are given few gifts in life and I threw that one away.

I felt rejected from a central part of my old life and not certain I was accepted in the uncharted territory of the middle classes. Even though I held high-profile jobs, establishment roles and supposed achievements never persuaded me to join it. My wife and I never networked, using our social life to build relationships that might be useful, and we preferred to make friends with people we liked, our neighbours, and to stay close to our family, especially my sisters and their children.

I suspect my own predisposition to independence, if not solitude, is what attracts me to Cuthbert. As I do, he had to deal with tensions in his life and upbringing. Perhaps he never resolved them, but at least he followed his own course, fought the demons in his heart and left high office at Old Melrose to try to know the mind of his God. That for me was a powerful motivation to follow in Cuthbert's wake, as he rowed downriver towards his destiny.

My broken rib had healed enough so that I could sleep, but my antics on the steep slope below Old Melrose had injured my back so badly that I found walking difficult, something that might be an inhibition later. I consulted a very practical and efficient chiropractor to help get me back on the road. As the years pile up behind me and those in front begin to diminish, I sometimes find myself musing on whether or not bits of me will simply cease to function, and in what order. Will my teeth last? Why has my bladder apparently shrunk to the size of a satsuma, and what is the evolutionary purpose of all that extra hair in my nostrils and ears?

In any case, I could not follow Cuthbert in my own curragh (and my wife had banned any other sort of craft; in fact, she was uncertain that I should be allowed out without supervision) and so I decided to spend time at places he would have passed on his journey, places of significance that might shine a brighter light on the long past. I thought I knew where he was going, and by the time we reached the secret tracts of a solitary place I hoped I would be able to walk in his footsteps.

As the Tweed wound its way north-eastwards, Cuthbert was rowing through the heart of turmoil and contradiction, a period of convulsive change all but ignored by Bede. Having expanded slowly from their stronghold on the rock at Bamburgh, the reach of the Anglian kings of Bernicia extended across much of north Northumberland in the second half of the sixth century. In Old Welsh, the language of the native British, *Bryneich* means something like 'the land between the hills' and as it was appropriated by the Bamburgh kings, it morphed into Bernicia. But not without a fight.

In the year 600 a host mustered in the fortress on Edinburgh's Castle Rock. These were the warriors of the kings of the Gododdin, and as they reined their swan-maned ponies south over the Lammermuirs their numbers grew in strength, as Cadrod, Lord of Calchvynydd, rode out to join them. Calchvynydd means 'chalk hill' and it is the oldest version of the name of Kelso on the banks of the Tweed.

After they forded the great river, the British cavalry plunged deep into what is now north Northumberland, Durham and north Yorkshire. They were searching for the warbands of the people they knew as Y Gynt, 'the Gentiles', the pagan Angles. They finally confronted their enemies at Catraeth, Catterick, on the River Swale, to decide not only who would rule over the north of Britain but also whether or not God's Christian soldiers would triumph over the heathens and idolators who faced them. For the British knew themselves as Y Bedydd, 'the Baptised'.

The Swale ran red with the blood of a great slaughter. On that terrible day, God was mocked, for it was the Anglian pagans who cut down the Baptised. Victory at Catterick was pivotal, for it soon won them control over vast swathes of territory, including much of the Tweed Valley, and their advance northwards was spearheaded by a remarkable general. Aethelfrith was known by his Old Welsh-speaking enemies as Am Fleisaur. It means 'the Trickster', or better, 'the Artful Dodger'. It seems that by stratagem as well as feral ferocity, Aethelfrith led his Anglian warbands to victory at the ancient fortress of Addinston in Upper Lauderdale and elsewhere. After 603, the Christian communities of the Tweed were in the hands of pagan overlords, the immediate ancestors of Cuthbert.

Between 1952 and 1962, the eminent archaeologist Brian Hope-Taylor discovered something of the nature of the paganism the Angles brought to the Tweed Basin. At Yeavering, now a farm with open fields on the the banks of the River Glen but then a royal centre in the foothills of the Cheviots, not far from Wooler, his excavators found the sole British example of a temple built to honour Anglo-Saxon gods. For the times, it was a large building, seventeen feet across by thirty-five feet long, and under its floor was a pit full of ox skulls and bones, the residue of sacrifice. Evidence from a contemporary burial in North Yorkshire suggested that in addition to cattle, human beings were also killed to propitiate the warlike pagan pantheon of the Angles. A nobleman had been placed in a grave and a living woman thrown on top of the corpse, pinned down and quickly covered with a mound of heavy stones. To prevent her screams from puncturing the solemnity of the ceremony, she may have been gagged.

In his treatise *On the Reckoning of Time*, Bede offered more information on the beliefs of his ancestors, even though his primary purpose was to establish a clear chronology for his magisterial *Ecclesiastical History of the English People*. The AD system of dating he used was invented by Dionysius Exiguus, an Eastern European monk who migrated to Rome in the sixth century. Calculating 1 AD as the year of Christ's birth, he created the Dionysian tables, which also worked out the correct date of Easter, accepted at the Synod of Whitby. But it was Bede's adoption of this system that established the AD and ultimately the BC method of reckoning time.

In his treatise some of the Anglian pagan festivals are listed. The most important seems to have been Modranect or Mothers' Night on 25 December, a clear antecedent of Christmas, while Blodmonath or Blood Month was November, when blood puddings were made after the slaughter of beasts before the onset of winter. Most dynasties in pagan England, including that of Aethelfrith, traced their ancestry from the paramount god, Woden, and he appears to have been closely linked to Thunor, the sky-god of lightning. Tiw was a war god and Friga a goddess of love. The Anglo-Saxon pantheon persisted long enough in Britain to name the days of the week and it seems likely that the conversion of pagans was a process rather than a series of events.

Sitting on the rowing bench of his curragh, no doubt allowing the current, the expression of God's will, to carry him downstream, Cuthbert moved through a landscape of great cultural complexity. The Baptised, the elite of native British society, had probably been converted by missionaries who walked over the watershed hills of the south-west of what is now Scotland, the Southern Uplands. In the century following the departure of the Roman provincial administration in 410, Carlisle still functioned as a small city complete with local government, and it was the centre of an early Christian parish, perhaps even a bishopric. Much later, in 685, when Cuthbert had become Bishop of Lindisfarne, he visited the city, and the royal reeve, Waga, gave him a tour of the Roman walls and, according to Bede, showed him 'a remarkable fountain that was built into them'. Some time around AD 400, a priest left the city to found a church in Galloway, Candida Casa, the Shining White House at Whithorn. Ninian was charged with a mission of conversion and, according to Bede, he preached the word of God to the Southern Picts.

Inscriptions on gravestones and elsewhere show Christianity advancing up Liddesdale and crossing the watershed hills to the Yarrow Valley and Peebles. Place-names add to the sense of early conversion. In Old Welsh, *eglwys* meant 'a church' and it is cognate to the Latin *ecclesia*. Ecclefechan in eastern Dumfriesshire is an ancient name and it means 'little church'. Much further east, Eccles in Berwickshire and Eccles Cairn near Yetholm in Roxburghshire were the sites of pre-Anglian churches, the former the centre of an early shire. By the sixth century, it is highly likely that there existed an organised ecclesiastical structure of some kind in the Tweed Valley. The impression given in Bede's work and the Anonymous *Life* that Cuthbert and other saints and monks were preaching to communities ignorant of Christ needs qualification. The native British, certainly their elites, had heard the word of God for a century and more before the invaders came, and it seems more likely that it was pagan Anglian settlers who were in need of salvation. Faint echoes of the prayers of British priests whisper along the banks of the Tweed. Perhaps Cuthbert heard them as he passed.

*   *   *

Two bats were feeding, swooping around the porch and the track when I took the dogs out on an early September dawn. Only the crows were croaking, like heavy smokers coughing when they wake up, and all the other birds were still roosting.

But once again, after only ten minutes and as the sky brightened, the little pipistrelles had crawled up and under the wooden fascia on the terrace to sleep for the day. It is not difficult to see how the Dracula association came about, but I like the wee bats and their flickering, frantic way of flying, making it very obvious even in the half dark that they are not birds.

On the eastern, leeward side, behind the stables, the old oak's leaves were turning brown and the young wood on the western side of the Bottom Track was opening up as its canopy began to fall. The rabbit and deer tunnels were much clearer and my little dog's sniffing showed where they led. If I had let her off the lead, I would never have seen her again. By the gate into the Deer Park track there is a stand of sweet poplars, and each spring and autumn they give off a powerful scent, slightly chemical like some women's perfume. It was strong this morning as the wee dog and I came back from checking the old horses, the mares and the mini Shetlands in the East Meadow. Bright in the east, with some shafts of weak sunlight, the sky was leaden over in the west and since that was where the wind was coming from we quickened our step to beat the band of rain. It moved through quickly.

There are different sorts of sunshine. This morning, the clouds streaked diagonally across the eastern horizon and the angle at which they moved allowed the western hills of Ettrick to be lit. Not strong or high enough to be dazzling, the sun made the heather glow and the greens of the upland pasture seem richer and more subtle. By the time the tideline had washed down the hills to our valley, the trees took on a luminous quality. It was perfectly still but with some moisture in the air; the land seemed to be still and peaceful.

I could hear no engine noise of any sort, only the lowing

of distant cows in the fields behind the Top Wood, the cawing of the crows strung out on the power lines and the songs of a few early birds. Sounds as well as sights are a transport to an older Scotland, a slower Scotland whose colours were glowing this morning. It seemed like a good time to get my walking boots on and, my sore back willing, follow the banks of the Tweed and the shadow of Cuthbert rowing through the river-mist for as long as I could.

When I reached Benrig, clouds had darkened the day again and rain was gathering in the south, over the heads of the hills. I parked by the gates to the cemetery and, having changed into rain gear, I followed a narrow path that snaked between a wall and a fence down which an avenue of mature trees marched. At the top of a high bank above the river, a steep, dark, damp and slippery set of wooden stairs led down to a narrow flood plain choked with hogweed and willow-herb.

There was little or no breeze, but in a moment, the weather changed once more, like a light switch. The sun suddenly lit one of the most beautiful reaches of the River Tweed I had yet seen. In my notebook, I see that I have breathlessly scribbled 'Stunning! So rich!' Downstream from Mertoun Bridge the Tweed is wide, with pattering shallows and a little flush of white water as the level of its bed suddenly drops close to the bank I was standing on, while on the far side the deeper pools swirled like liquid silk. I was standing below a heavily wooded high bank and on the other side biscuit-ripe cornfields led my eye eastwards. Just as there are below Mertoun Bridge, there are rows of ramrod-straight poplars in four ranks, their roots drinking deep from the constantly moist earth. The sun blinked between high clouds, alternately darkening the landscape and then flooding it with light, and the river glittered as it ran southeastwards to turn north again below Maxton Bank. I had been fretting about how sore my back and left leg would be after the scrapes and japes of climbing up to Old Melrose, but the glories of a stretch of the river I had never seen before drove all of that detail out of my mind.

As I walked on a good, if muddy path, I met only two dog walkers and reflected that in the seventh century, when Cuthbert rowed his curragh, I would have been standing on the banks of a busy highway. Fourteen centuries later, the shift has been dramatic. Only half a mile from the Tweed, the A699 carries all of the traffic now. I have driven that road many hundreds of times and yet had absolutely no sense of the secret beauties that flow so close to it.

Dug into the western bank above my path I noticed two stone archways closed by iron grills. At first I thought they might be limekilns, but a helpful bronze plaque explained that I was at the Crystal Well, or more correctly, a pumping station built to bring up the water from a lower well. Below the arches was a picturesque artificial grotto of the sort built by landscape gardeners in the eighteenth century and inside was a powerfully flowing little spring. I cupped my hands to taste sweet water and, as I bent down, noticed in the granite basin below the outflow that there were a few coins, something that is commonly seen in fountains and even below bridges over rivers. When people who throw coins into the water are asked why they do it, the general response is 'for luck'. But in fact what they are doing is a distant echo of an ancient ritual.

At Duddingston Loch, at the foot of Arthur's Seat in Edinburgh, archaeologists have recovered prehistoric swords, spear tips and a great deal of other military and domestic ironmongery from its silty, anaerobic depths. To the peoples of the first millennium BC and before, these were very precious, even prized items that appear to have been deliberately thrown into the water. At sites in the Borders and at watery places all over Britain, similar, smaller caches have been found. Historians believe that all of these objects were the deposit of ritual. Jetties were built out into lochs and rivers so that priestly figures could ceremonially cast these valuable swords, spears, shields and cauldrons into the water with no intent to retrieve them. These were in all likelihood acts of propitiation, of sacrifice to the water gods who lurked in the darkness of the depths, whose anger or ill nature could send flood and tide to play havoc.

Natural springs and wells also received deposits of metal, as well as other signs of reverence around them. Some years ago, I visited Madron Well in Cornwall, one of the most famous and most visited of these ancient springs. Tangled in a dense copse of thorn trees and other snagging undergrowth, the well was at first difficult to see as a single source, but its presence could not be mistaken as muddy streams trickled around the roots of the bushes. I eventually found what seemed to be the well, a spring surrounded by a set of squared-off kerbs, and tied on almost every available branch or twig around it were scraps of cloth, ribbons, small cuddly toys, a teddy bear and even bits of supermarket plastic bags. Some had messages or names on them, and I was told that many young girls and women come to make wishes at Madron. I saw no one attach any of these rather bedraggled, sad-looking favours, but I suspect the response as to why they did it would have been the same as the coin throwers', 'for luck', and maybe something to do with marriage or having children.

In fact, those who throw coins and other metal objects into water are also committing an act of propitiation, not so much for good luck as the avoidance of bad. And for me,

the discovery of the Crystal Well held infinitely more fascination than the elaborate construction of a pump to supply
water to the big house at the top of the bank, complete
with its plaque. I am certain the sweet water has bubbled
out of that bank for millennia and more, that it was venerated as a holy well long before there were monks at Old
Melrose. Water could be got from the Tweed, twenty yards
away, and in the many centuries before main drainage and
industrial effluent it would have been pristinely clear. But
wells like this had their own elemental magic; they were
spontaneous sources of the earth's bounty and places where
that could be blessed and celebrated. Perhaps Cuthbert
pulled his curragh over to the bank below the Crystal Well,
said a prayer and drank the holy water.

After spending time listening to the sounds of the river,
so peaceful they were almost hypnotic, I climbed back up
the slippery steps, the business of going up much easier than
coming down. The Pathfinder had plotted the ruins of St
Boisil's Chapel in Benrig Cemetery and on a knoll to the
south-east I could see an outline of very low walls, little
more than foundations. But before I reached it, I walked
through the rows of gravestones, noticing that several were
recent, one set up in 2016, and towards the edge of the high
bank I had climbed plenty of room had been left for more.
Closely cropped grass waited for the dead to be planted. In
my ignorance, I had imagined that because the old chapel
had been demolished in 1952, the cemetery would have closed
its gates to more funerals. But instead it seemed that families still wanted to bury their dead at Benrig, even though
there was no longer a kirk and the nearest lies about half a
mile south of the village of St Boswells. A small information
board explained that the Church of St Boswells (using the
modern spelling of Boisil) had been established in 1153 for

Lessudden, the old name of the village. But it did not explain why it had been built so far from where most of its parishioners lived. I suspect the attraction of such an inconvenient site was that the new church would be raised on ground that had been sacred for centuries, a place where saints walked. Perhaps the medieval church was built over a much earlier structure.

Places often seem not to lose their sanctity and sense of peace when formal worship ceases and I spent longer than I intended looking at the headstones, recognising one or two names, saddened by some that commemorated younger people, one a woman who had died in childbirth. Her baby had been buried with her and it reminded me of a prehistoric grave found in Denmark. A mother had probably also died in childbirth and her baby had been laid on a swan's wing and nestled in her embrace. Another headstone made me smile. Jeremy Church had died in 2000, only fifty-nine, and the inscription read: 'A witty, loving and unconventional man', with the second 'n' in unconventional reversed.

The demolition of the chapel dedicated to Cuthbert's soul-friend looked surprisingly complete. Only one or two courses of rough-hewn stone had been left above ground and some debris piled on top of those, but I noticed that several very old gravestones, their inscriptions long since faded, had been set into the old walls as the dead hugged close the sanctity of the ancient precinct, hoping their sins would be cleansed by the sacred earth. On the far side of the chapel were the higgledy-piggledy rows of much older stones, some lurching at drunken angles, others fallen. Around the southern edge of the cemetery were arranged the lairs of the local gentry, some of them surrounded by iron railings, setting them apart from the rest of us, even in death.

The path by the Tweed and the steep steps were part of St Cuthbert's Way and I rejoined it over a stile in the cemetery fence. It led through a dark wood dripping with last night's rain to more echoes of Old Melrose. Up on a mound, perched high over the river, stands the solid, foursquare block of Maxton Kirk. Dedicated to St Cuthbert, its existence has been noted in the written historical record for almost a thousand years and was probably a place of worship long before that. Maxton's location is suggestive of a pagan past. Old churches are often found on the sort of man-made mounds that may have formed the core of prehistoric monuments of an uncertain sort, perhaps burial places, perhaps the centre of a ditched perimeter. The dense wood of tall, mature broadleaf trees that stands between the kirk and the river mask how dramatic this eminence looked in the past. Standing high above the Tweed, it will have been visible from the north and west from long distances, what archaeologists call a statement in the landscape.

Unlike St Boisil's Chapel, and only half a mile distant, Maxton is still an active kirk. As I walked back to Benrig and my car, it struck me that this atmospheric reach of the Tweed and the churches and cemeteries on the high eastern bank are more than a long, lingering memory of the relationship between two saintly men, with their poignant dedications to Boisil and Cuthbert. I suspected that was a landscape already suffused with an older sanctity, and as I climbed over the stile back into the cemetery and walked over to my car, I could see Eildon Hill North through a gap in the trees, looking down benignly on this beautiful place. By the fence, the person who mowed the grass and kept the headstones tidy had piled the wreaths and spent flowers of recent ceremonies to rot on a green heap of clippings. But instead of reflecting that all is decay, I was warmed by and felt part of the continuity I found at Benrig and Maxton: the sense of continuing life in a place of death, as the twenty-first century still venerated what was believed to be holy by the uncounted generations of the past.

*   *   *

The swallows are leaving. Too intent on sheltering from the rain this morning, I did not notice until later that most of them have gone. Last night, they swirled around the stable yard and the house as usual, delighting us with their soaring and swooping, like the cadences of choral music. But this morning I could see none. Lindsay told me that there is still one pair feeding chicks in Blossom's box down in the stable yard. I fear for them, if they leave their immense, arduous journey to Southern Africa too late and hit bad weather with four young ones barely fledged. A friend told me that her swallows had all gone and she farms twelve miles south of here, in the Cheviot foothills.

It is a melancholy moment, the end of summer and winter soon to come. The house martins are also feeding a late brood and they are still showing off their awesome aerobatics, but in a short while they will fly south over the hills too. I was surprised that whatever impulse made the swallows begin their great journey took them at night – or perhaps they left at the very first peep of dawn. My knowledgeable friend told me they can fly 200 miles a day with a wind at their backs. Once our swallows cross the Cheviot Hills, they can be across the Channel in two or three days, before passing over western France, the Pyrenees, eastern Spain, Morocco and the Sahara. Over the wide sweep of the desert, there are few flies to be had and many die of starvation, especially young ones. But the survivors, the fittest, will overwinter in the warmth of Africa and come back to us next year. Swallows and martins mostly nest in buildings and I suspect they are tolerated, even loved, because they scoop up at least some of the pesky insects around the farm. I will miss them as we begin to hunker down for the bad weather.

It was a cold start, the first really autumnal morning, and out with my little dog I could see several trees beginning to turn. But the rain soon relented, and as the sky lifted I drove eastwards, following the Tweed, looking for Cuthbert. Having carved out its course after the end of the last ice age, the river hides itself in the landscape, only occasionally revealed from the road. As the rain spittered again on my windscreen, I was glad to have packed rain gear and put on waterproof boots. Out on the river in a curragh, Cuthbert would have had no such protection. In common with his fellow monks, he would have dressed simply in an undyed woollen robe or habit, and the story of the sea otters spoke of a loin-cloth worn under it. When it rained, Cuthbert would have been wet unless he had an early oilskin with him. These were animal skins smeared with fat or the oil from codfish and others to make them as waterproof as possible. On Inner Farne, the monks used the pig fat brought by the recalcitrant ravens to waterproof their boots. But in bad weather in the seventh century and long after, shelter would have been sought as quickly as possible when it rained. Cuthbert could easily have rowed to the banks of the river and turned over his curragh to keep himself dry until the skies cleared.

I wanted to walk beside the Tweed near a strange mound I had passed for many years but never stopped to climb. It looks simply plonked at the edge of a cornfield on the opposite bank of the Tweed from Makerstoun House. My dad used to repeat an old tale that its creation was the result of a dispute between rival landowners, the one on the south bank wanting to spoil the view from the windows of the big house on the north. I parked at a nearby farm and walked over a newly cut field dotted with round bales of straw as the sky began to lighten and the rain stopped. Skirting the mound, much bigger close-to than it looks from the road, I reached the high bank over the river and marvelled once more at its secluded, secret beauties. Below were a series of rapids over large shelves of exposed rock and beyond these a long, lazy stretch winding its way towards Kelso. In a grey rowing boat kept steady in the current by a ghillie, a woman sat on the swivel seat in the stern casting a long line over a pool between the rapids. In the prow sat two golden retrievers and they were the first to see me. The ghillie looked up and gave a friendly wave, as his companion focused on making good casts.

I wanted to get down to the riverbank to see if the mound could be seen from there, but a steep bank discouraged me. Still feeling the effects from my antics on the last one, I looked for somewhere less precipitous. Once on the river-bank, I found walking through the dense and lush vegetation very chancy, with several hidden drops causing problems, but it was certainly possible to see the mound from the river and I climbed gratefully back up to the cornfield.

The Law was steep, but not difficult to climb, and when I reached the top the view was disappointing. I could see little more of the river, and to the south only as far as a nearby ridge above Roxburgh Newtown Farm. But I did find something fascinating. The Law has a flat summit, perhaps as big as the footprint of an average house, and in one corner there is a strangely carved low pillar. It has three deep grooves cut on the top and some rectilinear markings on one side. Mobile phones are more than wonderfully handy cameras (I long ago abandoned taking paper notes in favour of photographing a record of walks and wanders): they can solve mysteries instantly. Having searched online for the Law at Makerstoun, I came up with the answer. This was a meridian pillar placed on the Law by a keen aristocratic astronomer and geophysicist, Sir Thomas Brisbane. He lived across the river at Makerstoun House in the first half of the nineteenth century. In order to make accurate readings of where true north (as opposed to magnetic north) lay, he had meridian pillars erected so that precise measurements could be taken. In an era when the British Empire and its attendant trade were expanding, this was more than a hobby or an interest.

But was the Law raised by Brisbane (who incidentally gave his name to the capital of Queensland) to site his meridian pillar on it? Other evidence, and what I could see, persuaded me that it was much older. The Royal Commission for Ancient and Historic Monuments recorded the discovery of terracing on the mound but does not hazard a conjecture about its purpose, agricultural or otherwise. Other sources remembered a holy well at the Law, one dedicated to St John. I could find no trace of any water source either on the top or around the foot of the mound. What I did notice was a distinct kerb all around the circumference. This will have been accentuated by annual ploughing, but I did see that in several places there was stone revetting. All that I could conclude was that this strange eminence had been man-made, and very probably long before Sir Thomas had workmen raise his meridian pillar on it.

Four huge beech trees have established themselves on the summit, growing to an immense height, their roots gripping the soil like thick fingers. They must post-date the period when the meridian stone was used because they stand between it and the observatory at Makerstoun, but they seemed to add something to the strangeness of this place. Carved on two of the more smooth-barked beeches were initials, one set dated 11 December 1922, almost a century old and now a considerable height off the ground, beyond the reach of contemporary knife-wielders. In the bole of one of these massive trees, water had been trapped where the roots knotted around each other, and to my amazement I saw two coins at the bottom of this tiny pool. On the north side of the mound, I came across something very puzzling: three canes had been rammed far enough into the ground to keep them upright in the wind, and on the end of each fluttered ragged flags that were so discoloured I could make nothing of them. Mystery is not confined to the unfathomable ceremonies of prehistory. For some people, their secret venerations on this old mound had enduring meaning.

If Cuthbert passed below the Law and prayed at the holy well dedicated to St John, it would have had some poignancy. When Boisil lay dying of the Yellow Plague, Cuthbert read him the Gospel of St John and they discussed it – as much as the old man's pain allowed. In the early medieval period, it was believed that the Book of Revelation was also written by the Apostle, although modern scholarship thinks that unlikely. Full of famous and extravagant imagery in its prophesies, Revelation culminates in the Second Coming of Christ, the time when God would walk once more in the Garden and Creation would be made anew. Many signs that the world to come is at hand are found in the book that is also known as the Apocalypse, derived from a Greek word that means not catastrophic disaster but an unveiling, and these include outbreaks of plague. After Boisil's death, it may be that Cuthbert believed the end times were coming and he fled

Old Melrose so that he could both become closer to God and pray for his own salvation and also that of others. His journey down the Tweed may have been an urgent mission.

Making my way carefully down off the Law, I was fascinated to watch a cloud of tiny birds, finches, I think, playing in the bushy young trees on the northern slopes. They seemed little bigger than butterflies, and moved skittishly across the high willowherb and the dried crackling stalks of cow parsley, not staying long enough on each stem, it seemed to me, to eat the seeds. Perhaps fifty of them, they sometimes all flew in harmony, so that for a moment the sun caught the bright plumage of their bellies. The cloud of little birds suddenly glinted like blown sunlight and then they seemed to disappear as they perched once more on the stalks of the tall plants and were still. It was mesmerising. Autumn may be a melancholy farewell to summer but it still holds small moments of quiet beauty.

On the way back down to my car, these feeding birds sparked a memory of childhood. I noticed that rosehips were ripening in the hedgerow. When I was a little boy, we used to pick the red hips and take bags of them to school to be weighed. The Delrosa company paid about 3d a pound, I think, and from them they made rosehip syrup, a sweet cordial rich in vitamin C that was spooned into the unwilling mouths of children. We were given cards to mark and, for a week or two, evenings were taken up as groups of kids stripped the prickly bushes by the roadsides and in the woods. It was fun, but surely a fragile business model, one with no chance of working now, what with health, safety, and Uncle Tom Cobley and all. I haven't seen a bottle of Delrosa for years, or anyone gathering the hips. It seems like a waste; there were plenty of bushes hung with thick bunches of their red fruit after the long, hot summer.

Having hurt my back clambering up steep slopes, I was pleased to feel less pain on this brief expedition. I thought I could begin walking again for some longer distances so that I might follow Cuthbert downriver as he fled from the world.

Amongst the first blooms of spring, primroses were my mother's favourite and most years she took my sisters and I to Daniel's Den. On a wooded bank by the Tweed that shelved steeply downwards like the biblical pit, many clumps of these fresh little yellow flowers grew and we took as many bunches home as we could carry. They seemed not to flourish for long indoors, but their picking was a welcome ritual, an expedition that made my mother smile, and it was a smile that made us smile.

On my way down to the riverbank, I passed the old wood, now putting on its autumn colours, and began walking along the edge of history, my own as well as Scotland's. Rearing up from the floodplain of the Tweed, and with the River Teviot guarding its eastern flank, are the relics of an ancient fortress, what was once a stockade, a stone-built castle and for me a crucible of dreams. What we knew as the Old Castle when we were little boys is marked on the Ordnance Survey as Roxburgh Castle. It sits on top of a vast oblong kaim or mound piled up partly by geology and very substantially by human hand. A few shattered fragments of the medieval castle's mighty walls survive, impossibly thick, teetering on the edge of the steep slopes.

As a child, I was fascinated by the Old Castle. Even though the site is obscured and choked by many mature trees and chest-high nettles and willowherb, it was a place where my imagination was fired. I could reconstruct the massive gate-houses, the keep, and fire arrows or roll down boulders or pour hot oil on any who dared attack the mighty fortress. I knew all the stories of its sieges and invented more of my own. I convinced myself there were tunnels under the rivers, that the kaim was hollow and that there was treasure buried – somewhere.

The site has never been excavated, and as a young teenager I persuaded my friends that we should borrow garden trowels and dig for history, maybe even treasure. Instead we found fragments of green and brown glazed medieval pots, uncovered a flagstone pavement only just under the grass near the north-east gatehouse and came across a great deal of fleetingly exciting rubbish. Half-buried under neglect, spectacularly situated, the commanding and protecting presence above Roxburgh, Scotland's first recorded town, now entirely effaced, not one stone left standing on another, this rich, mysterious place made me a historian. As I climbed up to where we had scrabbled around fifty-five years ago, much of that excitement came flooding back. I even kicked around the loose dirt with my boot, vaguely wondering what I might uncover.

In those far-off summer holidays, all that interested me was the medieval castle and how it had changed hands often, sometimes garrisoned by Scots, other times by English soldiers. It was only much later that I realised it had a longer and even more intriguing history. In 1999, after I had left the world of television, Weidenfeld and Nicolson published *Arthur and the Lost Kingdoms*, in which I made a historical case (one that still persuades me) that Roxburgh Castle was once a base for the shadowy post-Roman warlord. The old name was Marchidun. In Old Welsh, it means 'horse fort' and I argued that Arthur led hosts of cavalry warriors against those who attacked and invaded the faded Roman province of Britannia: Anglo-Saxons, Picts and others.

Some time in the early seventh century, almost certainly after Aethelfrith's victory at Addinston in 603, an Anglian warrior know as Hroc took control of the old horse fort, after Arthur was long dead and the armies of the native kingdoms everywhere defeated. In common with other warlike cultures, Anglian chieftains often adopted animal names, and Hroc means 'the rook'. On top of the vast kaim he had a burh built, a fortified stockade, probably on the highest point, where the medieval keep later stood. The entire area, which had been encircled by the walls of the medieval castle, would have been an impossibly long perimeter to garrison. By the time Cuthbert rowed his curragh down the wide reach of the Tweed to the north-west, this fascinating place had become Hroc's burh, later rubbed smooth into Roxburgh. Perhaps sentries on the palisade saw him pass.

Across the busy road that skirts the deep ditch of the moat on the same side, a tree-lined track leads down to the river and a fishermen's hut. I wanted to walk the wide sweep of the haughland that leads around a loop to the town of Kelso, my home place. Grazed by sheep and cattle, this broad expanse leads up to a knoll which was marked on old maps as High Town. It was the centre of Roxburgh, where urban life began in Scotland in the early twelfth century. Documents in the cartulary of Kelso Abbey, founded in 1128, detail the names of streets – King's Street, Market Street and the Headgate – two churches, a Franciscan friary, a mint, a school and a bustling market. All of it has completely disappeared, leaving no visible trace of any kind. With the coming of the Wars of Independence in the late thirteenth century, the wool trade that built Roxburgh shrivelled and gradually the town sank into oblivion beneath the grass. The castle was eventually destroyed, 'doung to the ground' by the Scots in the late fifteenth century, in case it fell into English hands.

I walked up to a stand of copper beeches and wondered

at the ruthless transit of history. As late as the 1290s I would have been standing at the corner of a busy market place where wool merchants from as far afield as Italy bargained with producers, many of them monks or their agents. The four great Border abbeys at Melrose, Kelso, Dryburgh and Jedburgh all ran vast sheep ranches in the Cheviots and the Lammermuirs, and their annual wool crop generated so much cash that the Abbot of Melrose was once able to borrow against it. And yet instead of the bustle of commerce, the creak of cartwheels, the smell of fresh bread from street ovens and the stink of tanning pits, only a gentle breeze blew up from the river, riffling the long grass in the autumn sunshine. I had brought a small rucksack with me and it was warm enough to sit down by the bole of a wide chestnut tree to eat my cheese sandwiches and wash them down with bottled water. Across the haughland and a wide reach of the Tweed I could see the Cobby, a stretch of riverbank where local people were permitted to fish and swim. The name remembers a time before bridges and derives from 'coble', a raft-like ferry that once plied between the two banks on a diagonal course to allow for the current.

In the 1950s and early 1960s, Kelso pumped its untreated sewage into the Tweed at the Cobby, and yet no one thought anything of swimming and paddling in the filthy water. Sometimes children's skin was badly affected with what we called water blobs: strange, almost translucent blisters. Hard by the Duke's Dyke – a vast and high encircling wall built around Floors Castle and its wide policies in the early nineteenth century by local people in a project paid for by the duke to provide income for unemployed and destitute families – there used to be a diving board. Below was a deep, mostly natural pool that lay upstream from the effluent, and I can remember my dad launching himself off the top board.

That terrified me, even though he was a good swimmer. When a severe stroke almost killed him at the age of fifty-eight and deprived him of movement in one arm and partially in one leg, he was still able to swim in the indoor pool that had been built in Kelso by then.

I walked downhill towards the river, dipping further to where a ditch had once run around the town of Roxburgh. This part of what is now called Friars Haugh (after a Franciscan friary built south of the town, beyond the walls) encompasses all of the pasture down to the point where the River Teviot joins the Tweed at the Junction Pool, making it a wide, stately, even majestic river. The pool is an expensive and productive place to fish and a grey rowing boat is often seen eddying in the conjoined currents. About a hundred yards upstream is the cauld, a breakwater originally built by the monks of Kelso Abbey and their lay workers to create a lade that would direct the flow of the river to the wheels of their mill. Part of an ancient arch survives, and the deep channel disappears into a dangerous darkness under it. Remarkably, a modern corn mill, run by John Hogarth Ltd, is built right on top of the Abbey Mill and its mighty pantechnicons still transport the finest oatmeal and pearl barley in the world, a near thousand-year continuity.

Built up to divert the current, the cauld (maddeningly noted as a 'weir' on the otherwise excellent Pathfinder maps) had the effect of creating a short fall of white water, as the Tweed spilled over its edges. In addition, it has enabled the build-up of small, oblong river islands called the annas. A term unique to the banks of the Tweed, it may derive from the Gaelic *annaid* meaning 'church lands', perhaps a distant memory of Aidan and the Irish monks at Old Melrose. Growing up in the 1950s and early 1960s, the annas were a dangerous and exotic playground. Trees, hogweed and dense undergrowth, all made lush by the river and the silt it swept down, had grown up, and winter spates deposited all sorts of debris on these islets, sometimes whole tree trunks, roots and all. Once we came across part of a tractor cab, a tangle of old signposts and much else that could be turned over and examined by little boys who had waded the river at the Cobby.

Our principal summer holiday sport was rat hunting. The annas were infested with them and, feeding on the corn from the adjacent mill, some rats were the size of cats. If cornered, they were likely to turn and fly at any attacker. Armed with whuppie sticks, stout staves cut from trees on the annas, we thrashed the undergrowth to flush them out. I remember seeing a huge rat trying to escape into the river, its back end as big as a football and its orange tail long and whip-like. Looking back, it was a crazy, cruel and pointless thing to do, its only purpose to generate excitement and tall stories.

To Cuthbert's eye, the river will have looked very different as it looped around the peninsula Kelso is built on. Under the modern town may lie the fleeting whispers of the wooden cells of a diseart like those in the loops of the Tweed at Dryburgh and Old Melrose, but all that is known of the site before the foundation of the abbey in 1128 is that there was a pre-existing church dedicated to St Mary. And while the loop of the river is not as extravagant as at Old Melrose, there exist the remains of a strip of boggy ground to the north. It may be where an older course of the Tweed ran and could have served as a barrier or a perimeter for a sacred precinct. However that may be, Kelso will always be hallowed ground for me.

As Cuthbert fled from the world, guiding his curragh around the Maxwheel, a powerful, treacherous tangle of currents where the Tweed is turned north-east by a high river cliff, his mind will have filled with thoughts of his mission. Early Christians often acted consciously in imitation of Christ, following his actions and examples literally. When David I of Scotland invited communities of monks northwards to found his Border abbeys, thirteen came, their abbot and twelve brothers, like Jesus and the Apostles. Cuthbert's faith was leading him to a different life from the communities of monasteries; he was seeking the secret tracts of a solitary place, but also following Jesus' example. After his baptism by John, Christ entered the desert 'with only wild animals' for company, and there he endured the temptations placed before him by Satan. Fasting, struggling against the pangs of hunger, he resisted the offer to turn stones into loaves of bread, uttering the much quoted 'One cannot live by bread alone', followed by the much less quoted 'but by every word that proceeds from the word of God'. When his faith and trust was tested by Satan's challenge to jump from the 'pinnacle of the temple', Christ replied, 'You shall not put the Lord, your God, to the test'. Faith and not certainty was what mattered. In the final temptation, on a high mountain where 'all the kingdoms of the world' could be seen, Christ rejected the offer of power and the adulation of crowds.

Cuthbert was steeped in these stories from the gospels and could recite them from memory when he preached. He believed in their literal as well as spiritual truth. The Anonymous *Life* was explicit, saying that the saint 'fled from worldly glory', rejecting it just as Christ had done. And in further emulation he sought solitary places just as Christ had done when he wandered into the trackless desert – and did so for the same reason. Devils lurked there. Places where no one lived were the resort of an army of demons, all of them very real to Cuthbert and his contemporaries, and he sought to do battle with them and their temptations. Many were the fallen angels brought down by Lucifer. Molech, Chemosh and Samael all appear in the Old Testament. In Revelation, Abaddon was the King of the Abyss, Asmodai the Demon of Wrath and Mammon the Demon of Avarice. The desert was haunted by the Goat-Demon, Azazel, and by the enormity of Behemoth. Some of these have survived as metaphors in modern culture, but the existence of the Devil and other malign beings has faded out of fashion in recent times. We no longer fear them, or even think about them. But for Cuthbert their hellish power was real and, as he fasted and prayed, he aimed to drive back Satan and his evil angels into the shadows where they belonged. In the deserts he would find, Cuthbert would use the power of prayer to call down God's help and vanquish the demonic legions not only for the sake of his own salvation but also to save other souls.

Guided he hoped by God, Cuthbert allowed the currents of the great river to bear him eastwards, further and further from the cares of the world he had left at Old Melrose, towards the rising sun. Little more than a mile downstream from Kelso, on a promontory north of the farm at Whitmuirhaugh stood a place the saint may have wished to avoid. Aerial photographs of fertile riverside cornfields have shown up traces of a large seventh-century settlement otherwise invisible from the ground, its wooden buildings long since rotted and disappeared. It was enclosed by a ditch and a stockade rammed into the upcast on the landward side and defended by the wide river to the west. The outline of large rectangular timber halls, a possible church, partly sunken houses known as grubenhauser, and a field system complete with drainage ditching can all be seen as cropmarks or discolourations of the topsoil, and the photography is so clear that almost every one of 368 graves in an adjacent cemetery can be made out.

Whitmuirhaugh was probably what Bede called an *urbs regis*, a royal township, and as such will have been visited by the kings of Northumbria as they progressed around their realm. They apparently arrived in some pomp, a procession preceded by the royal standard and also a post-Roman emblem known as a *tufa*. A winged orb set on a pole, it may have been carried in front of the king as the symbol of the Bretwalda, the Britain-ruler, a title several Northumbrians claimed. When smoke from the cooking fires from the houses and halls at Whitmuirhaugh came into view, as Cuthbert rounded the bend of the Tweed, he may have kept rowing, not wishing to explain himself to the royal prefect who governed this large and now all but invisible settlement.

A few hundred yards downstream the little River Eden flows into the Tweed and it may be that Cuthbert rowed over to the left bank and pulled his curragh some way up its course so that he could visit a place where he might pray, a place already sacred. Close to Edenmouth Bridge I parked my car and put on waterproof walking boots so that I too could make a detour, but into my own past as well as the vanished world of the seventh century. Climbing fences and staying as close to the Eden as I could, I found it well named. Its little valley is indeed Edenic, very beautiful, quiet, still and small, a place shaped by the hands and sweat of men and women who cultivated the fields that are moulded to the meandering of the river and its gentle slopes on either side. Many admire, even revere the wildness and the majesty of the mountains of the Highlands, but for me it is the landscapes that people have made and cared for that I love. They are memories of uncounted generations who grew food and tended animals in the fields and on the hillsides, and the lower Eden valley is a lovely palimpsest of all that day-in, day-out labour. The lives of kings, queens, saints and the notorious are recorded in history books, visitor centres, street names and in a host of other ways, but until the coming of the census in the middle of the nineteenth century, the voices of others are largely silent. The fields, however, seem to me to remember their people, those who tended them. Until recent times, they had no other monument.

I noticed that at the end of this very dry summer the Eden was low, certainly too shallow for even a light craft like a curragh. But no matter, neither Cuthbert nor I had far to go.

The tiny village of Ednam is ancient, beginning life with the Anglian name of Edenham, and it is the oldest parish to come on record in Scotland, the subject of a grant from King Edgar in 1105 to a man called Thor Longus. Probably of Scandinavian origin, Thor the Tall was given Ednam in return for services rendered to the king, probably military. The charter noted that the land was '*deserta*'. This may mean that no one farmed there, but given the fertility of its free-draining fields and their helpful southern orientation, I think that would be surprising. It may be that the charter makes an early reference to a former diseart, also a possible translation of '*deserta*'. The Eden does loop at Ednam, but the site looks unlikely. Perhaps a holy man, his name long forgotten, lived there in a simple wooden cell or hermitage. Encouragement for the notion of the village as an early Christian focus of some sort is supplied by a bell. In the Royal Museum of Scotland, the Ednam handbell is preserved and it is very old indeed. Probably used to summon a congregation much as church bells in a tower or steeple do now (and also possibly used for cursing – apostates, sinners or even exorcising devils), it was cast from iron some time in the seventh century. Its clapper has not survived, but when rung it will have made a distinctive noise. In a landscape where only natural sounds were heard, the harsh clang of a bell would have made people look up.

The old kirk at Ednam is beautiful – and dedicated to St Cuthbert. The entrance to the kirkyard is also the way into a private house and there are notices advising visitors on the subject of parking. Not that they are much needed, except perhaps on Sundays. The day I came there was no one about, and in any case I had left my car at the mouth of the Eden. Part of my purpose was to revisit an extra-ordinary coincidence.

Bina Moffat, my grandmother, was born in 1890 at Cliftonhill Farm, close to the village. On my way to the kirk, I had walked through its bottom fields. In the kirkyard, there is a tall headstone with the names of my direct ancestors carved on it. For a farm worker paid in kind as well as cash, it is a surprisingly expensive stone and I am sure it was an expression of great love and loss, for it was erected by my great-great-grandfather, William Moffat, in memory of his wife, Margaret Jaffrey. She died at Cliftonhill Farm in 1891, a year after my grannie, Bina, was born. Margaret was only sixty-four. Into the same lair, William Moffat's coffin was lowered some time after 1 March 1896, when he died at Wormerlaw Farm, about two miles north-west of Ednam. Their daughters came to join them in 1920 and 1931.

I have often thought about these people I never knew and their lives at Cliftonhill, Wormerlaw and other farm places. They made my grannie, and she made my dad, who grew up in Kelso with his great-aunts. And my grannie helped raise me, as my mum went out to work at the local Co-op and did night shifts in what would now be called a care home. Working on the land – what Gran called the 'auld life' – they were out in most weathers without waterproofs, sacking across their shoulders to keep out the worst of the rain, bending their backs in the fields and the stackyard, coming back to their cottage when the light failed, tired and hungry. Life was harsh, certainly, and without free health care they died younger than we do now.

Farmers paid part of their workers' fees in what were called 'gains' – potatoes, oatmeal, coal and other consumables – and, despite the elaborate headstone in the kirkyard, there was little money. When Bina was born in 1890, William Moffat, his wife Margaret and their three daughters (including Annie, my great-grandmother) all lived in a two-room cottage. In such a small space, their lives were necessarily intimate, sharing beds and sitting around the warmth of the fire and the range as winter winds whistled up the Eden Valley.

In those decades of high farming, ploughmen and other farm labourers often moved, but rarely very far. They were employed for fixed terms at hiring fairs held every six months in local towns. As farm workers stood in groups in the market square or street, farmers struck bargains with a handshake, a coin and sometimes a drink in the local public house. But in truth, these fairs could be a bitter experience, and publicly shaming for those whose hands were not shaken and were left standing at the end of the day, unemployed, unable to feed their families. William Moffat was first horseman, or head ploughman, at Cliftonhill and he was an attractive man to hire because he could bring three unmarried daughters into the bargain as supplementary field workers. Bina was an illegitimate child, as was my dad, and so there was no husband to claim Annie. In the farms of the late nineteenth and early twentieth centuries, 'wumman workers' did all the jobs not associated with horses, such as weeding, singling, shawing, harvesting root crops, dairy work and sometimes scything. My grannie remembered her mother and her aunts singing as they worked in the fields around Cliftonhill, moving along in a line, the music helping to keep a rhythm.

In an age before mass transport, life for my family was local. They went to church down the hill at Ednam, celebrated in the stackyard at harvest time, danced in the village hall and went courting only as far as they could walk. In Annie's case, more than courting. Over 250 years, as far back as the census and parish records allowed me to trace their lives, my family did not move far, living out their hard-working lives little more than two or three miles from Kelso, where I was born, and on the farms of western Berwickshire.

Sometimes I remember to take flowers to Ednam kirkyard, and it is a place that always seems peaceful, settled, even a sanctuary when problems pile up. I have no sense of my ancestors, these people I never knew, raging in their graves. They lie quiet, I think, at peace after all those long days in the outbye fields. There are no photographs of them, but I believe I can hear them, their voices whispering across the furrows and pastures of the farms of the Eden Valley. My dead may be invisible to me but they are never absent, and one day when she is old enough to remember if not under-stand, I will take Grace and any other grandchildren we might be blessed with to Ednam kirkyard and tell her some-thing of where she came from, who her people are.

By contrast with me, my wife has led a very far-travelled life. The daughter of a regular soldier, Lindsay was born in Hamburg, lived for a time in Hong Kong and went to boarding school in Gloucestershire. Sometimes, she reflects that there was nowhere she could call home – until we came to live in Edinburgh for twenty-five years, and then to the Borders for another seventeen. And so it was astonishing to discover that less than twenty yards from the headstone William Moffat had erected in Ednam kirkyard there is another impressive memorial that commemorates her direct ancestor, a tenant farmer from the mouth of the Eden called Robert Kerss. He died in 1849 when William Moffat had begun work as a ploughman in the farms around Ednam. They may have met, known each other or of each other. It is such an astonishing coincidence that I do not know what to make of it, except to observe that I am not the only one to have come back home to the Scottish Borders.

The Tweed flows through a history of disharmony; on its banks stand the blackened ruins of centuries of war, the forgotten altars of a disputed sanctity, and again and again memories of division peep through the dense under-growth. As he rowed on eastwards, Cuthbert will have been unaware of all of this, for it lay far in the future. He would live and die in the kingdom of Northumbria as it rose from a nest of pirates on Bamburgh Castle Rock to become a beacon in Western Europe that glowed with the creativity of Bede of Jarrow, the raising of the gorgeously painted and carved crosses at Bewcastle and Ruthwell, the writing and decorating of the *Lindisfarne Gospels*, and the forging and working of jewellery of glittering opulence. Over four centuries, Northumbria forged a common culture between the Firth of Forth and the River Tees and beyond. The Scots language arrived as a dialect of Northumbrian English and early patterns of place-naming and land management were established in the Lothians and the Tweed Basin. And most significantly, the genetic mix of the communities of old Northumbria formed the basis of the modern population.

Not far beyond the place where the Eden joins the Tweed, the river makes yet another long, lazy loop around Birgham Haugh, and as I made my way along the southern bank early on a bright September morning I eventually arrived at the cause of centuries of disharmony, times when fire and plunder came repeatedly to the Scottish Borders, when rivers of blood ran and great slaughter stained the pages of history. Where the river reaches its most southerly point before meeting the North Sea at Berwick, an insignificant little burn trickles into the Tweed. Known as the Reddenburn, it has formed the border between England and Scotland for at least eight hundred years. Even at its mouth I found that I could easily step across it, hopping from one grassy bank to another, moving between two different national jurisdictions. And yet despite the fact that this little stream draws a thick, dangerous, black line through our history, I find myself able to straddle it in all senses. My ancestral DNA was Northumbrian long before my nationality was defined as Scottish, and my people, the men and women who worked the fertile fields around Ednam, are the same people who farm the land to the east of the Reddenburn, people who must call themselves English. The price of that distinction has been paid in blood, the sacrifice and waste of hundreds of thousands of lives.

Perhaps only a few hundred yards into England lies Carham. It existed in Bede's time and he recorded the name as Aet Carrum. In 1018 this sleepy little hamlet would ring to the clash of steel. Malcolm II, king of much of what would become Scotland, made an alliance with Owain, the last king of Strathclyde, and in May of that year, in the woods of the narrow valley of the Caddon Water, between Galashiels and Selkirk, they massed their host for the descent into the old and failing kingdom of Northumbria. As the war-smoke billowed from their campfires, Malcolm, Owain and their captains laid their battle plans.

Scouts reported that Earl Eadulf of Bamburgh (the old royal title had long since been downgraded by the ascendant Wessex kings) had reached the Tweed at the head of a force of spearmen but that it was not large. Malcolm and Owain hurried down the Tweed to meet them and the hosts clashed on the haughland near Carham. On 26 May 1018, the grass was soaked with blood, as Malcolm's Highlanders roared their war-cries, incited each other into '*freagarrachan*', or 'rage-fits', and swung their axes as they pushed back Eadulf's shield-wall. Triumphant, the Highlanders climbed over the wrack of dying and screaming men and drove the defeated Northumbrians into the river. The chronicler, Symeon of Durham, recorded great carnage at Carham when he wrote 'all the people who dwelt between Tees and Tweed were well-nigh exterminated'.

To the men and women who stood up on the ridges to the south to watch the battle rage below them, the Scots did not defeat the English. In their reality, a Gaelic-speaking king and his Highlanders had invaded Northumbria, their homeland, and slaughtered their people. Like the warriors who died on the banks of the river, they spoke English, and when it became clear that Malcolm II's host would prevail, they would have fled in fear of their lives and their homes.

Having sought permission from the vicar, I climbed the decidedly rickety stairs inside the bell tower of the old parish church at Carham so that I could take photographs of the river as it turned towards the sea. To the north, I could see the line of the main street of the village of Birgham, or at least the back of the houses on its south side. It lies in Scotland because from the Reddenburn the frontier follows the midstream of the Tweed almost to Berwick. The original place-name may have been Briggaham and a bridge may have crossed the river nearby. But the village claims a corner of history for another reason. Perhaps on the park called the Treaty Field, a group of Scottish barons gathered to meet English envoys in 1290. They may have chosen Birgham because there was a bridge. Four years earlier, Alexander III had been killed when his horse lost its way in the stormy darkness and both of them fell to their deaths over the cliffs near Kinghorn on the Firth of Forth. His heir was his granddaughter, a seven-year-old Norwegian princess. Known as the Maid of Norway, Margaret became Queen of Scotland, and almost immediately diplomacy began to decide her future.

Edward I had conquered Wales in the previous decade but he realised that his designs on Scotland would be better served by dynastic marriage than the spilling of more blood, to say nothing of the expense. At Birgham a nervous arrangement was agreed by the Scots. Edward of Caernarfon, the heir to England, was betrothed to Margaret of Norway, but the Scots baulked at the crippling conditions Edward I's negotiators attempted to impose, such as the surrender of Roxburgh and Berwick castles and the acceptance of the virtual overlordship of Anthony Bek, the Prince-Bishop of Durham, a man who styled himself the heir of Cuthbert. In the event, the little girl died in Orkney on her way to Scotland, and claimants to the vacant Scottish crown quickly began to contend. This process eventually resolved into the failed kingship of John Baliol and his deposition led to the Wars of Independence. When he was stripped of the crown at Stracathro, near Montrose, Edward I is said to have remarked, 'A man does good business when he rids himself of a turd.' For almost three centuries, and with few respites, armies marched across the fields of the Tweed Valley, trailing destruction behind them. From my vantage point at Carham, I could see the innocuous little village where the history of the Borders began to unravel.

Creation looked well that morning, man's as well as God's. And by God's I mean the transit of the seasons and all of the shifts of the natural world we cannot influence. The sun lit the riffling leaves and the lush grass as the birds welcomed the morning, and a fresh autumn breeze blew along the river. On the edges of the woods near Carham Hall elder-berries were growing in abundance. Heavy clusters of these beautiful purple berries hung from the trees, their weight bending the branches low. Bina, my grannie, used to make elderberry wine and with my sisters, Barbara and Marjie, I was sent out to pick clusters – with very specific instructions. I can't remember what they were, just Bina's wagging finger. I imagine she told us how to recognise ripe berries and leave the green ones. When she had enough to fill her big jam pan (it was called the enamel pan, dark blue on the outside, white on the inside, and seemed like a small bath), she began the ritual. Using a potato masher, my grannie crushed the berries after we had turned our hands blue stripping them off the stalks. Then she boiled it all up with water before tipping in bags of sugar. To get the mixture to ferment, she made toast, a thick slice of near-burnt toast. On it was slathered yeast and it floated on the surface of the blue-black mixture. I am not sure, but this may have been an old farm-cottage way of slowly introducing the yeast as the toast soaked and it seeped down. When we were coughing with colds, Bina used to give Barbara, Marjie and me thimblefuls of elderberry wine and it was sweet and rich, like port – alcohol for the under fives. I am certain my mum had no idea and we never told her.

In the distance I could see the high, grassy mound of

Wark Castle rising above the bank of the river. Much smaller than Roxburgh but almost as dramatic, it glowers over the Tweed, standing guard on the frontier, keeping watch over the fields of Berwickshire and Scotland across the water. Sheep grazed its flanks, following the sun around them, and on the road to the east of the castle a tractor trundled downhill, a trailer bouncing behind it. In early June 1314, this quiet landscape would have looked very different.

To finish his father's work, complete the conquest of Scotland and defeat the traitor Robert Bruce, Edward II mustered an army at Wark. On the flat haughland, almost eighteen thousand men had answered the royal call to arms and marched or ridden north. Latecomers saw a vast military camp sprawling westwards along the banks of the Tweed for more than four miles, reaching at least as far as Carham. In the midst of a myriad of pavilions, awnings, temporary corrals, parked ox-carts and the spiralling smoke of cooking fires Wark Castle rose on its mound and the royal standards fluttered from its walls. Between eight and ten thousand animals were devouring the summer grass for miles around it. The much-feared English armoured knights had brought their heavy horses, their destriers, and there were perhaps two and a half thousand of these valuable creatures being managed by their grooms, grazing in their halters, drinking deep from the Tweed. And there were even more riding horses, ponies, mules and oxen, as well as herds of cattle and flocks of sheep, a mobile food supply.

Once the muster was complete, the great army descended to the river, splashed across the fords of Wark and entered the realm of Scotland. Over the heads of the spearmen, the men at arms, archers and armoured knights, many banners fluttered, but one of these was thought to have great talismanic power. The Prince-Bishop of Durham, Richard Kellaw, had brought with him the precious Banner of St Cuthbert and all believed that it would bring victory against the barbarian, godless Scots. In the event, the English army was routed at Bannockburn in 1314. But Cuthbert had brought victory almost two centuries before, when in 1138 David I's invading army had been defeated near Northallerton in Yorkshire. Mounted on a battle cart, the presence of the holy banners from Durham and the minsters at York, Ripon and Beverly, was thought to have been so powerful in a decisive English victory that it became known as the Battle of the Standard. The Banner of St Cuthbert was also carried at the disastrous defeat of the Scots at Flodden in 1513, only four or five miles from where I walked.

Soldiers fought more fiercely because they believed Cuthbert was by their sides, in a physical sense, and that his certain presence meant God too was with them. The banner was not large. Square and mounted in loops on a horizontal pole, it was made principally from red silk and velvet. Crucially, part of the shroud of Cuthbert had been sewn into its centre behind white velvet that had been embroidered with a cross. It is not clear from the sixteenth-century description that this was the silk cloth given to Cuthbert by the Abbess Verca when he visited her nunnery on Coquet Island off the Northumberland coast, but what mattered was that the army believed it was. And what is equally striking is the denial of history. By the early twelfth century Cuthbert had become an emphatically English saint, perhaps one of their most powerful weapons, despite the fact that he had been born and raised north of the Tweed and in Northumbria. And this saintly, fragile and peaceful man would have been appalled at the way he had been recruited into an English army intent on slaughter. The Englishing of Cuthbert would run like a nationalist ribbon along the frontier for centuries.

More irony waited further along the riverbank. East of Wark lies an international anomaly. Plotted on my trusty Pathfinder, and also on the most recent Ordnance Survey maps, such as the spectacularly unwieldy Explorer series, is a kink, a shift of the frontier from the midstream of the Tweed. As the river bends around Lees Haugh, south of the town of Coldstream, the line abruptly dips south into England and brings two or three acres back into Scotland. No one is sure why.

However, nineteenth-century histories repeat an entertaining tradition. This small parcel of Scotland used to be known as the Ba' Green and it was said that each year the men of Coldstream would play an early, anarchic version of rugby against the men of Wark. The prize was this piece of England. But when Coldstream grew much larger than Wark, the annual fixture became so one-sided that it was eventually abandoned – with Scotland keeping the Ba' Green. It is cheering to see the august and precise Ordnance Survey remember such an unlikely tradition – but what other explanation fits this remarkable oddity?

* * *

On my way down the Tweed, travelling in the shadow of Cuthbert, I had walked in stages of only short distances, not trusting my back and my wonky left leg to carry me for more than two or three miles at a time, and preferably on the flat ground of the riverbank. I decided it was time to take on a longer journey and I pulled out my rucksack and packed it. Frankly, I was becoming more than a little frustrated by all of my aches and pains. They themselves were becoming a pain.

My life seems increasingly governed by the relative severity of all my aches, as my body copes less and less well even with quotidian demands, things I could do with perfect ease a year or two ago. At sixty-eight I am not ancient (well, not very ancient), but my body seems to be losing function faster than I can think of workarounds, and this baleful process has been accelerated by the miserable and seemingly interminable business first of reconstructive surgery to fix my shoulders in 2017, and now with my back and leg. Pain bores a hole in your brain, and its more or less constant presence is not only unpleasant, but is also making me so much less productive than I used to be. I refuse to be stoical, meekly accepting. Sod it. I need to fight, block it out and bloody well get on with things.

I have always had a strong work ethic, linked to my non-PC sense of myself as a provider for my family. I also feel that food and drink cannot really be relished unless they have been earned, and these days I am not doing enough, frankly, to deserve an evening by the fire with a glass of something good. That's if I can manage to carry the bottles from the shop to the car without yelping at the soreness in my dodgy shoulder.

I realise that this is a dour and dismal value system. Maybe it is genetic. My Ednam ancestors all worked on the land through scores of winters, enduring the cold and wet (without waterproofs), and they wore out their bodies. By the time they had reached their sixties (if they were lucky), they were din, Scots for 'done', meaning worn out and weakened. When he was fifty-eight, my dad had two massive strokes, and while my mum waited by the phone for news from the Western General in Edinburgh I went down the street in Kelso to do some basic shopping. I met an old lady well known for her directness and she asked for news. 'Aye,' she sighed, 'it's no' as if he was a din man.' Perhaps that's what I am becoming: a din man.

But that sounds like an excuse: the top of the slippery slope of using age as a qualifier: not bad for sixty, sixty-five, seventy . . . Who says? Apart from being made grumpy by pain, my brain is no more clouded than it has ever been. And there is no reason why I can't still do good work, none at all.

Fortified by a refusal to be pathetic, I pulled out my rucksack, made some cheese sandwiches, wrapped them in foil and packed some standard kit – dry socks, maps, compass, waterproof, hat, water, charged-up mobile phone and a notebook. In the early morning I would drive downriver, park the car and follow Cuthbert as he sought the secret tracts of solitude, places where he would pray and make a covenant with silence.

6

The Quiet Waters By

Two recent forecasts of high winds had not amounted to much, and as I put my rucksack in the back of the car early on a September morning, it was breezy, but nothing alarming. However, an instinct sent me to check the BBC News website, where worrying reports were coming in from the west, Northern Ireland and Galloway of gale force winds, and they seemed to be heading in our direction. At their predicted arrival time in the Borders, nothing changed. Maybe the wind had turned in another direction. Nevertheless, something told me to wait.

In the middle of the morning, the breeze began to freshen. I saw very few birds in the sky, as clouds scudded eastwards at high altitude. And Lindsay decided to bring in all the horses. Minutes after she shut the last of the stable doors, the storm raced over the hills to the west, driving heavy rain in front of it, and smashed into us. Suddenly a gale was raging.

At noon, Lindsay came rushing into my office. 'Come and look!' What? 'Just look!' Our glorious, lustrous, purple-leaf maple tree that stood by the gates to the stable yard had been torn in half by the storm and its main trunk completely blocked the track, hanging from a huge gash. It was a stunning, majestic tree, but its wide maple leaves were its downfall. As the eighty-mile-an-hour gale blew directly at it, the leaves acted like thousands of postcard size sails and the trunk had been ripped down its middle. In shock at the suddenness of such destruction, we gawped for a moment at the death of a tree we had planted, one we had watched grow and had come to love.

But it had to be destroyed even more completely, and quickly. I ran to alert my son, Adam, to fetch the chainsaw. In the teeth of driving rain that stung us like spears of ice, we dismembered the fallen tree, cutting its smooth, healthy limbs to pieces, dragging the branches and their fatal leaves off the track and jamming them into the back of the pick-up so that the storm could not pick them up and smash them into buildings or worse.

In the evening, when the storm had calmed to a strong breeze, Lindsay and I tried to count our blessings; no human beings, no animals and no buildings (apart from fences) had come to harm. But in truth we were dumbstruck, grief-stricken at the loss of an old friend. As she came up from the stables at the end of each day, having given the horses their night-time haynets, Lindsay told me she sometimes said goodnight to our beautiful tree. Now it is goodbye, after only a few moments of elemental rage. But thank goodness no one was injured.

The day after the storm dawned bright and only breezy; it seemed a good day to be walking with Cuthbert. Our track is so pock-marked with countless puddles and ruts, it forces a very slow descent to the tarmac of the lane at the bottom. I have shredded too many tyres to rush, and crawling along in first gear has encouraged me to look up and notice more and more detail, the transit of subtle changes from day to day. This morning the last few swallows were gathering after the storm, a process known as staging, and they lined up on the telephone cable beside the track like crochets on a music score. I counted twenty, some of them very small, all of them the parents and broods from the nests in the stable yard. Some of the wee ones looked impossibly fragile, and for all those hatched and fledged on our farm this summer their immense journey to Africa will test their endurance and will be unlike anything else they have experienced. Perhaps they fly in the slipstream of their parents. As I reached the lane, I wondered if the storm had been a signal.

When I parked at Twizel Bridge an hour later, I noticed a large flock of swallows wheeling over the fields by the side of the road. High in the sky, occasionally swooping low over the river, they seemed to be getting ready for the flight south, perhaps gauging the strength and direction of the wind, sensing the temperature, feeling the air in some unknowable way. What prompts all of this concerted activity both mystifies and awes me. I love the swallows and count the winter months until they return.

Twizel Bridge crosses the River Till, a tributary of the Tweed, and history has marched across it for centuries. Completed in 1511, it was the longest single-arch bridge in Britain until 1727, reaching ninety feet across the river. Two years after the masons had paved it, the vanguard of the English army tramped across Twizel Bridge to outflank James IV of Scotland's forces. They had encamped on Branxton Hill at Flodden, well to the south. Ten thousand men trundled the cannon behind them that would rake the ranks of Scottish spearmen a few hours later, as a crushing defeat was inflicted and James IV was killed. Behind the English battle lines, the Banner of St Cuthbert was flown, its red silk fluttering above the ruck of hand-to-hand fighting.

A modern bridge now carries the traffic between the Scottish Borders and Berwick-upon-Tweed, but Twizel Bridge is still open to walkers, and from its parapet I looked east down the Till. When Cuthbert rowed his curragh to the place where it joins the Tweed, I believe that he pulled on the left oar and turned it up the tributary. This eventually leads close to the moors and mosses above Ford and Doddington, to the crags and caves of the Kyloe Hills, to places that were remote and trackless, where he might retreat completely from the world and find peace to pray, and by fasting and self-denial bring himself to a transcendent state where he might better know the mind of God.

On the Ordnance Survey, the Till looks a small river, meandering below the eastern foothills of the Cheviots, but what first struck me on that bright morning was how deep and in places how wide it was. And even when the sun shone and the bankside trees glistened, the river was not so much picturesque as dramatic, even sinister in places. River names are amongst the oldest and least changed in the landscape, so ancient that their derivation can only be guessed at. Till is obscure, but it shares an initial letter with many of the major rivers of Britain that empty into the North Sea. The T-rivers include the Thames, the Trent, Tees, Tyne, Tweed and Tay. It is a coincidence that may be more than surprising and might make a thread-like link to a lost language. The derivation of Tay is thought to be Taus, and that probably in turn relates to the Sanskrit word *tavas*, which means 'to surge'. Variations in the Indo-European languages related to Sanskrit might account for the similarities in these ancient river names.

Walking along the banks of the Till, I felt the sense of *genius loci* was powerful, and I quickly came to understand the sentiment behind a rhyme that has survived in local folklore and that I knew as a child:

Tweed said to Till
'What gars ye rin sae still?'
Says Till to Tweed,

'Though ye rin wi' speed
And I rin slaw
Whar ye droon yin man
I droon twa'.

Chilling. And as it turned out, a worthwhile warning to
those like myself who spend time looking around and not
where they are going. The Ordnance Survey showed a path
all the way from Twizel Bridge to the village of Etal, but I
wanted to begin where Cuthbert had begun his journey into
the moors and hills – at the confluence of the Till with the
Tweed. It proved impossible to walk up the southern bank
of the river and so I was forced to make a detour to St
Cuthbert's Farm, so that I could follow the man of God
from there.

It was by now a magical early autumn morning. The wind
had dropped and I'd soon stuffed my body warmer into my
rucksack. Till meets Tweed at an anna, a river island called
Little Haugh, and the currents of the rivers run so slow and
deep that no movement is perceptible. A swan glided out
from the bank of the anna and slowly made its way to the
mouth of the Till. It seemed like a good omen.

The map also marks the site of St Cuthbert's Chapel
(which had probably given the farm its name) and it was
sad to see the state it was in. Marooned in the midst of a
ploughed field, it was like a metaphor for the decline and
isolation of religion. Roofless, the rectangular structure was
medieval, built in 1311 on the footprint of a much older
chapel. Its arches had been prevented from collapsing by
some shoring brickwork and the weeds and small trees that
had grown inside had been cut down and treated with weed-
killer of some sort, but otherwise it was in a neglected,
lonely state. The chapel perches above the confluence of

Till and Tweed, and it will have been visible from both rivers, but its history is less obvious, difficult to untangle from the legends that grew up around Cuthbert.

Viking raids on Britain began with an attack on Lindisfarne in 793, and by the early ninth century, or perhaps later, the monks felt forced to abandon their vulnerable coastal church. They took Cuthbert's body and his relics with them to begin their wanderings over southern Scotland and northern England, and legends grew up quickly. In his long poem *Marmion*, Walter Scott repeated one of the more unlikely tales. Having sought refuge at Old Melrose, he wrote, the brothers took Cuthbert's body down the Tweed in a stone boat. It foundered at the mouth of the Till, where for centuries something that looked a bit like a stone boat lay on the bank of the river. Perhaps it was a natural rock formation that encouraged the tale. Offshore from Dunbar in East Lothian, St Baldred's Boat is in fact a dangerous rocky outcrop. Above the confluence of the rivers a chapel was built and in the early Middle Ages it may have served a small village called Tillmouth, all trace of which has been obliterated.

To me, on that blessed September morning, when the wind had calmed after the storm, the little chapel seemed less like a relic of ancient miracles or even a metaphor. It was more of a signpost that told me to turn and follow the Till, so that I could stay close to Cuthbert as he fled from temptation to seek the solitude of the moors and hills.

At first the going was gratifyingly easy, as a metalled road followed the eastern bank of the river past two Hansel and Gretel cottages that looked as if they had been built by the wealthy owner of an estate as decoration as much as accommodation. It would be like living in a cuckoo clock. As I marched on, any thoughts of my back and dodgy leg were banished by curiosity. The river surprised me: its pools were dark and deep and in places it looked fifty yards across. Cuthbert would have had no difficulty navigating the shallow draught of his curragh up the Till. Where it was occasionally shallow and rocky, he could easily have kilted up his robes and waded, towing the boat and whatever he had stowed in it for his new and solitary life. Soon I came across the first of a long series of sheer sandstone river cliffs that had been scarted out by the glaciers of the last ice age as the young river found its course. It occurred to me that hermits liked cliffs and the occasional caves that could be found where the strata folded or fractured. Just before Twizel Mill and its complex of channels or lades, built to direct the water to its wheels, I saw a heron standing still as a stick on a rock, searching for movement in the pool below. A friend once memorably described herons as Presbyterian flamingos.

When the track turned uphill behind the mill, it abruptly changed character. From a well-made road through leafy north Northumberland, I found myself suddenly thrashing through a temperate version of the Amazonian jungle, stung repeatedly by weapons-grade nettles and having to stop every few yards to work out where the path might go. Almost as bothersome were the repeated mouthfuls of spider webs as I brushed past tall undergrowth. At one point I confirmed I was on the right track because I saw a fresh bootprint on a molehill. But if moles were happily digging under a track, it might not be, well, a track. It was very difficult going indeed, and where low branches were too thick to push aside I had to duck low under them or occasionally crawl, hanking my rucksack as I did so. I should have packed a machete.

Eventually, I emerged from the tangled fringe of unkempt woodland by the river and came to what looked like the policies of a big house, perhaps Heaton Hill House. Hallelujah! At the other end of a new but very narrow wooden bridge, clearly intended to restrict passage across a deep burn to walkers only, I came upon more architecture by Hansel and Gretel, or possibly Bilbo Baggins. Clearly new and raised up off the ground with logs, a dainty, pale-blue, one-room hexagonal hut with an arched doorway, small windows and a chimney that turned out to have been built to house a barbecue. Set at table height, a fire pit was surrounded by benches covered with what looked like deer-skins and cushions. For barbecues in winter or bad weather, the hut looked an excellent idea, if more than a little twee.

Wide and green haughland lay beyond and, grateful for liberation from nettle stings, insect bites and facefuls of spider webs, I marched on along the riverbank. In the shade of a huge beech tree of the sort often found in the policies of big houses, a fisherman stood midstream in the Till, elegantly and expertly casting and recasting a luminous green line. The colour made the arc of his cast clearly visible. At my cautious and quiet approach, he reeled in and waded over to talk. Hoping for salmon, he had been even more pleased to get a sea-trout earlier and told me in what I judged to be a Geordie accent how much he loved the river and its peace. Having been advised by him that there was a path by the riverbank, I carried on past a spectacular, sheer sand-stone cliff. It was like an illustration from a geology textbook of how sedimentary rock is formed. At the bottom were thick, hard folded strata while at the top, fifty feet and millions of years away, was a lighter coloured sandy soil that was being compressed, infinitely slowly, into harder rock. On top of that there was a thin layer of grass. Very neat. On a flat face at the bottom of the cliff, someone had etched a stylised tree, perhaps the tree of life, and near it there was an overhang that might well have sheltered a river traveller on a wet night in the seventh century.

But there was no path. It simply ran out in a dead stop at the foot of the cliffs. Instead of my usual instinct for rashness, I turned back. And for some distance, as many steps were retraced. I then followed a more elevated path, looking down on the river as it bent around a neatly situated house surrounded by beautiful trees (all seemed to have survived the gale, possibly because their location was so sheltered by the woods on the high riverbank) and closely clipped lawns. But then this path petered out and forced yet another retreat, the third of the day. You would think that after many miles walked I would have some idea by now how to do this properly. Rather than Ranulph Fiennes, this was exploration by Groucho Marx. Minutes later, I came across a low signpost that forgave a little of my waywardness. Placed some distance from the river, it read 'Riverbank Path eroded, use Higher Footpath'. Well, thank you very much.

By the middle of the morning, it had grown even warmer, and between my rucksack and my back my shirt was damp. The Higher Path turned to be ill-cared-for and frankly dangerous, especially for the likes of me, with my propensity for pratfalls. Having clambered over a rotted and partially collapsed stile (where I hanked a bootlace on an exposed nail and twisted my leg), I followed a path that veered very close to another high cliff that dropped sixty feet sheer to the river.

Eventually, I thrashed out of the woods to meet a wide, made track with a very discouraging signpost. It told me that in two hours I had come one and three-quarter miles from Twizel Bridge and it was four miles to Etal. Pathetic. Disconcertingly, the track led sharply downhill and turned back the way I had come. When it reached the riverbank, however, it began to follow the course of the Till south-east towards the village. I saw signs with enigmatic numbers on them and realised that they were the reason for the good road. They marked different fishing beats, and fishermen had to get themselves and their kit to them by car. I made good time and by twelve noon, the country lunch hour for those who breakfast at seven in the morning, I sat down on a log to eat my sandwiches – smoked salmon, not cheese, a homage to the river.

I sat for longer than I meant to, watching ducks feeding, alerted by honking to a high arrowhead of geese flying south, and taking time to settle down after all the scratches, bites and scrambles from my morning in a temperate jungle. The peace of Cuthbert at last descended and I reflected that the elapse of fourteen centuries meant a great deal even in quiet, green and apparently unchanging places like this.

As a historian I have always tried, as far as is possible, to put myself in the shoes of people from our past, to imagine the landscape they inhabited. Wood was a vital resource for everyone in Britain until the Industrial Revolution began to produce synthetic materials in great volume. It was needed to build houses, make tools, carts and boats, and for heating and cooking. Bark made buckets and other containers, while specific woods like ash and alder were fashioned into spears and shields. That dependence made the seventh-century landscape different. Cuthbert rowed up a river that would not have been lined with mature woodland planted by estate owners for pleasure and good looks. In the early medieval period, and likely before that, trees were farmed and often zealously protected with rights, such as the gathering of firewood and the pasturing of pigs on beechmast carefully circumscribed. And in the same way, the rights to fishing and the pursuit of wildlife such as ducks, swans, geese and all of the other fauna that lived on the river, were defined in great detail. But while it looked different, and many more people will have lived on its banks, the Till would still have been peaceful and beautiful.

After making my way across more open haughland, I came across another strip of riverside woodland. The Ordnance Survey clearly showed a path through it, but when I came to a gate, a blue sign indicated in capital letters 'RIVERSIDE STRICTLY PRIVATE' and an arrow pointed left to direct walkers to a public footpath that led into a large gorse bush. In Scotland I would have had no difficulty ignoring this sign because of the freedoms of the right to roam legislation and the much older sense in Scots law that there was no such thing as trespass. But this was England, home of shotgun-toting retired military types, faces puce with rage, eyes bulging, moustaches twitching, bulldogs snarling. So I opened the gate and carried on. The Ordnance Survey is infallible, mostly, is it not?

After less than five minutes in Colonel Mustard's Wood, I heard gunshots. Close by. And stopped. I recognised the softer pop of shotgun cartridges, a contrast with the sharper report of a rifle. I had seen several blue drums, those used as pheasant feeders, in the woods I had come through and then also remembered, better late than never, that the game-shooting season might be underway. Having paused to look carefully up the densely wooded bank, I could see no movement or hear any dogs or beaters putting up birds. Nevertheless, I found I was suddenly walking a lot more briskly. And my anxiety was not reduced when I came across a bench with a game bag on it and no one around who might own it. Eventually a path led diagonally up the bank and I found a very well-made track shelved into the rising ground. Beside it was a sign indicating a footpath by the riverbank that led to Twizel, exactly where I thought I had been walking. But it also carried a sign for a house I thought

I had yet to reach. I looked again at the OS map and realised that I had mistaken one riverside wood for another and taken the wrong route. There was indeed a path which the blue sign prohibited, except that it was fairly overgrown – and definitely private. This was England, and I had been trespassing, and the Colonel would have been well within his rights to shoot me. By accident, of course. Or at least set the bulldogs on me. Sometimes I think my wife is right. I should not be allowed out on my own.

Staying well away from the edge, I followed the road until it arrived at one of the reasons I wanted to walk down the banks of the Till, and why I believed Cuthbert had chosen this route. The Ordnance Survey marks the site as St Mary's Chapel but gives little idea of what a strange place it was to build a church. Little more than one or two courses of foundations, the ruins lie on a shelf long ago scooped out from the high and steep riverbank. Measuring only about fifteen by thirty feet, the old floor area was mostly puddled with water that had not fallen from the sky but bubbled up from a spring across the track. Beside it stood two ramshackle corrugated-iron sheds that had a wood-encased pipe connecting them with the bankside, the place where the spring surfaced. All that signified an enduring sanctity was a small modern cross placed on a low plinth, probably where the altar once stood. The spring water, probably holy in early medieval eyes, also had a more modern function, one I could not discover. Perhaps it had become a mineral spring.

St Mary's Chapel intrigued me because I surmised that Cuthbert may have known of it, or at least the holy well that seeped out of the riverbank beside it. The dedication is also suggestive. Many of the great churches of the Tweed Valley, Melrose and Kelso abbeys amongst them, were dedicated to St Mary and my research had prompted the notion of a local cult around the life of the Virgin. This odd little church above the Till may have been part of this early group of sacred places. I could find no way to get down to the remains of the church – at least no way that looked safe – and so I walked on another few hundred yards to the pretty village of Etal, the other reason that drew me down the river with Cuthbert. Etal derives from Eata's Haugh, and this may well be the clearest link of all, for it was the name of the Abbot of Old Melrose who agreed to admit Cuthbert and who later became the Abbot of Lindisfarne.

I am almost certain that Cuthbert visited Etal, and in the company of Eata at a time when neither were at peace with the world. Some time in the late 650s, King Aldfrith of Northumbria invited the Melrose abbot to found a new monastery at Ripon in Yorkshire. Eata took Cuthbert south and appointed him Guestmaster, the first important role for the young monk. But political tensions in the Church between the Irish Celtic strain of Christianity and those who believed that Roman practices and leadership should prevail were approaching a crisis and they forced a radical reversal of royal policy. When Eata refused to accept the direction of Rome and the papacy, he and Cuthbert were forced out of Ripon. This was done at the instigation of Wilfred, precisely the sort of churchman Cuthbert did not wish to be. Having travelled twice on pilgrimage to Rome, Wilfred was a deeply political figure who moved in and out of favour with the Northumbrian kings, sometimes at the centre of power, at other times sent into exile. He employed a retinue of armed retainers, and when the Melrose monks were ejected from Ripon he immediately took over and by 664 found himself on the right side of history, as the Synod of Whitby ruled in his favour.

Wilfred and Cuthbert were rivals in another sense. Soon

after his death on Inner Farne in 687, a saintly cult swirled around the memories and miracles of Cuthbert's exemplary life. When copies of the Anonymous biography began to circulate after 705, Wilfred was in exile, and two years before, he and his supporters had been excommunicated by a council summoned by Aldfrith. When the king died in 705, Wilfred returned even more powerful as the adoptive father of Osred, Aldfrith's successor on the throne of Northumbria. Before his own death in 709, this highly political priest was in control of the bishopric of Hexham and of the monastery at Lindisfarne. Compiled to bribe and influence, his treasure was divided amongst his followers and very quickly they commissioned a *Life* of Wilfred that related miracles and promoted his own saintly cult. Written by Stephanus Eddius, a choirmaster brought north from Kent, the *Vita Wilfridis* was copied and recopied. When it began to circulate, Bede of Jarrow set to work to counter it with his own prose *Life of Cuthbert*. The lives and works of these two saints could not have been more different.

At the end of the seventh century, and for some considerable time afterwards, the conferring of sainthood was largely a local matter, and if miracles were associated with a monk, nun or priest, or indeed a secular figure (such as King Oswald of Northumbria), then that was usually enough. When lives of saints, listing their miracles, acts of charity and examples of their piety, appeared, reputations were consolidated and cults began to flourish.

By the middle of the twelfth century, Pope Alexander II decided that the papacy should have the exclusive right to make saints. The process instituted and refined by medieval popes has changed very little. It begins with the spontaneous growth of a posthumous cult around the life and works of an outstandingly pious person, often a priest, nun, bishop or even a pope. An application is then made to the Congregation for the Causes of Saints, and the Vatican bureaucracy rumbles into action. Sometimes it can take centuries. A miracle is needed, at least one, and supporters of the cult figure often try to bring one about by praying that he or she will intercede with God. Virtually any incident thought to be related to the prayers, such as a recovery from a serious or life-threatening illness, is then investigated by the Congregation. If the miracle is substantiated, then the cult figure advances to a different stage and becomes beatified. Such demi-saints are given the title of Blessed. To move on from there to sainthood, a second miracle is needed, and if that bears examination, then a case is prepared to go before the Congregation court. A Postulator argues that sainthood ought to be conferred and a figure known as the Devil's Advocate argues against. If the case is proven, only then are saints made.

History has not been kind to one saintly woman Cuthbert knew well. Queen Aethelthryth, or Etheldreda, the wife of King Ecgfrith of Northumbria, had a chapel dedicated to her in Yetholmshire in the foothills of the Cheviots, one of Lindisfarne's major possessions. Under the French rendering of her name, St Audrey became very popular in the south of England in the medieval period. Held on her feast day and birthday, St Audrey's Fair at Ely became famous, particularly for the fine quality of the jewellery and silk scarves on sale. Sadly, these were supplanted over time by cheap imitations and the word 'tawdry' was coined as a contraction of St Audrey.

When Eata and Cuthbert set out on the long and melancholy journey from Ripon back to Old Melrose, they probably travelled north on the old Roman road of Dere Street. Still in good condition, it lay close to the monastery seized by Wilfred's supporters, perhaps his armed retinue. When the road crossed Hadrian's Wall, the two companions may have continued northwards on the Devil's Causeway, another Roman road that led to the port at the mouth of the Tweed at Berwick. It crosses the Till some way upstream from Etal, and the fastest and most sensible route from there to Old Melrose was by curragh, allowing the current to take them down the river to its confluence with the Tweed. A larger boat, rowed by several oarsmen, could easily have made good time as Eata and Cuthbert returned to their mother monastery. After their expulsion from Ripon, it must have seemed like a sanctuary.

Apart from St Mary's Chapel, no trace of these ancient associations remains above ground. Instead, Etal is a single street of white cottages, some of them thatched and all refurbished by the owner of the local estate in the early twentieth century. The street leads to a medieval castle. Even though it is only half a dozen miles from the Scottish border, Etal looks very English indeed, a Miss Marple village that would not look out of place in the Cotswolds or the Chilterns.

Cuthbert had an unwitting hand in the beginnings of these stark cultural differences because his fame and exemplary life gave rise to one of the earliest definitions of Englishness. Much of early Northumbria was organised into shires, usually much smaller areas than the more modern counties that were largely abolished in 1974. Some of these little shires came into the possession of Lindisfarne and the collective term for the church's estates became St Cuthbert's Land, a label later used by the prince-bishops of Durham to describe their vast holdings in the north of England. Islandshire stretched along the coastal plain from south of Lindisfarne up to the mouth of the Tweed. The farms of this fertile area supplied most of the monks' needs and their produce was carted across the sands at low tide. The Till flows through Norhamshire and its centre was the *matrix ecclesia*, the mother church at Norham, a place that would come to play a part in the creation of the cult of Cuthbert.

When the border between England and Scotland began to harden along the banks of the Tweed, some of Lindisfarne's possessions, mainly gifts from the king and the nobility of old Northumbria, were cut adrift. Coldinghamshire, the estates around the monastery where Cuthbert had his feet dried by the sea-otters, found itself in an anomalous position since it lay well to the north of the border. Yetholmshire in the foothills of the Cheviots was eventually divided up. In a process that is difficult to date securely but was underway by the twelfth century, the people who lived in Islandshire and Norhamshire began to call themselves the Haliwerfolc, the people of the Holy Man, the people of Cuthbert. The glories and memories of Northumbria were fading as the border between England and Scotland began to fracture a culture fashioned and shared on both sides of the Tweed.

But there turned out to be compensations. Like all the best English villages, Etal has an excellent tearoom. Tired after thrashing my way along the Till, and at least three hours since my salmon sandwiches, I sat down gratefully with a pot of tea, a fruit scone and small jars of clotted cream and raspberry jam. Even though the waitress had microwaved the scone to warm it up, thus rendering its texture indistinguishable from foam rubber, it was all delightful and very welcome.

A warm afternoon was becoming a beautiful evening and, refreshed by an hour in the garden of the tearoom, I decided to leave the banks of the Till and walk eastwards. Not far from Etal, and down a shady B road, lies another model village. After 1860, Ford was almost entirely rebuilt by its owner,

Lady Waterford, as a memorial to her husband, who had died in a riding accident. In front of a group of substantial houses and a well-made village hall, the widow had a dramatic monument erected. On a smooth, marble column a large and slightly menacing angel stands, his wings folded behind him in such a way as to reach the same height as his head. I felt he might swoop down at any moment, like a celestial Batman.

Ford Castle is much older than the village and much rebuilt. It played a central role in perhaps the greatest historical drama to play out in this quiet and pretty part of north Northumberland. In September 1513, James IV mustered what may have been the largest army ever to march out of Scotland. Allied to France against England, the young king saw an opportunity. Henry VIII's army had taken the towns of Tournai and Therouanne, and the French implored James 'to advance a yard into England' and open a second front. Perhaps 30,000 soldiers forded the Tweed at Norham, took the castle of the prince-bishops and moved south to Etal and Ford. Their castles surrendered and James IV made his headquarters at Ford, where it was said he spent too much time in the company of its keeper, the beautiful Lady Heron.

Meanwhile the Earl of Surrey raised an army in the northern counties, taking the Banner of St Cuthbert from Durham Cathedral as a rallying point. His forces mustered at Alnwick and a curious diplomatic exchange began. Couched in the elaborate language of chivalry, Surrey sent the Rougemont Pursuivant herald to the Scottish camp at Ford with a message. With great politeness, the English general enquired if the Scottish king was prepared to do battle on 9 September and were the hours between 12 noon and 3 p.m. convenient? The Islay herald was despatched to the English camp with an equally polite message, saying that these arrangements were agreeable.

On the morning of 9 September, James IV led the Scots from Ford Castle and they crossed the Till to occupy a fortified hilltop position at Flodden. Lookouts saw the English begin to advance from the south-east. Marching in three battalions, they halted, formed up on the Milfield Plain and pitched camp at Barmoor, an elevated site that offered good visibility in all directions. At that point Surrey learned of the heavily defended position of the Scots at Flodden and was not pleased. He sent the Rougemont Pursuivant with a less polite message. His proposal was that the armies should fight on open ground, on the Milfield Plain. But James IV would have none of it; he refused to allow the English herald into his presence and sent a message to Surrey he was a king and would do as he pleased. At that moment, all of the niceties of late medieval diplomacy were jettisoned and military realities took over.

The English army then began to move, but not towards the Scottish position. Instead they marched north to the Tweed and for some time James IV thought they were leaving the field. But before they reached the river the English suddenly swung westwards and crossed the Till, the vanguard and the artillery rumbled over the new bridge at Twizel, while the bulk of the army splashed over at Straw Ford, the place where I had been eating my sandwiches. The seventy-year-old Earl of Surrey was outflanking and out-thinking the young Scottish king by leading his army north of his position, thereby cutting off any possibility of retreat back to Scotland. The old warrior must have been confident.

As soon as his manoeuvres were complete, Surrey saw an opportunity. Using his more flexible and more sophisticated field artillery, his gunners began to rake the Scottish ranks, and the more ponderous Scottish siege guns could make no effective reply. James IV had no option but to advance from his position of strength on Branxton Hill. And then everything went disastrously wrong.

Positioned in the van of his men, perhaps near the front ranks, charging on foot, the Scottish king ran down the hill and tore into the thick of the English halberdiers. The Spanish ambassador, Pedro de Ayala, later wrote 'he is not a good captain because he begins to fight before he has given his orders'.

But James's immense physical courage almost succeeded because his battalion pushed, hacked and bludgeoned to within a spear's length of breaking through the English lines, reaching very close to the position of the Earl of Surrey. He sat on horseback in the rear ranks so that he could see how the fighting ebbed and flowed and give orders to reinforce, pull back or attack. As the battle descended into a melee of desperate hand-to-hand fighting and the ruck of battle closed in, James was on foot and in no position to command, unable to see further than the desperate struggles around him. After some hours of exhausting combat, the king's battalion was rolled up and surrounded and, with many of his noblemen, James was killed.

On my way home from Ford and a day on the banks of the Till, I drove to the site of the battle, the place where five hundred years before almost fifteen thousand men died. And for many it was a slow, agonising death. Blows from bladed weapons rarely killed outright and many will have lain for hours badly wounded on the battlefield, passing in and out of consciousness, slowly bleeding to death. There is a monument close to where the battle was fought, but nothing more to remember all of the terrible slaughter that took place. As the sun dipped behind the Cheviots, I found that my imagination failed me. All I could do was bless the historical accident that my generation had not been forced to face a war and all that shocking, savage waste of life.

When I arrived home, these dark thoughts instantly fled. There was some late sun and Kim had brought out our granddaughter Grace for a walk before bedtime. The wee lass could not get outside enough and her parents had had to buy her a sunhat for the summer of summers. We went down to the stable yard to see the horses, sleepy after their hard feed, and then over to some shrubs where the bees were still busy. They fascinated her. Hearing the distant drone of engines, we looked up to see an aeroplane flying south-east, its silver undercarriage brilliantly lit by the low sun. When I told Grace to look up, she saw it, and after a moment the wee girl waved. I am not at all sure why, but I felt tears come and hastily blew my nose. It may have been the impossible innocence, and the scale. A tiny toddler holding her grandpa's hand and waving at a huge machine thundering across the face of the earth.

Our farm is a place full of memory, the deposit of centuries of experience, but it also lives in the present and will change in the future. The landscape and its animals make demands, and these force Lindsay and I to cope and adapt. With Grace, it will be different. She is the first child of this farm for many years, and through her eyes we will see it fresh and new. I look forward to her racing around with the dogs (the term 'dodos' coined by Grace will be cast aside as baby talk, used by no one except me), helping with the daily chores, sitting next to me in the pick-up and telling me her story of this place. I hope there will be years enough for it to be a long story.

7

Wandering

The following morning I set out for the wild up-country of north Northumberland, where I believe Cuthbert went to open his heart to God, where he would pray and keep silent, listening for whispers of revelation on the ever-present wind. The maps of Doddington and Ford moors are speckled with memories of the saint, some of them long-standing and more persistent than mere tradition. Cuddy was an affectionate nickname for Cuthbert and on that bright morning I set out to find Cuddy's Cave, a place where he is said to have prayed and sought refuge.

The course of the Till led me to a very striking outcrop of rock on the flanks of Dod Law, where the land rose steeply and the moorland began. Poking out of the hillside, it looked like the corner tower of a vast submerged stone fortress. After my usual difficulties with high barbed-wire fences, I climbed up the steep slope to be astonished. The vista to the west was vast, completely unexpected. I could make out landmarks that were many miles distant. Across the Milfield Plain, where the Earl of Surrey had preferred to do battle with James IV, the eastern ranges of the Cheviots petered out in the low hills above Wooler, and due west, the conspicuous rounded hump of Yeavering Bell stood out clearly against the horizon. Beyond that, the hills faded in the haze of a sunny morning. When

Cuthbert came across this great rock, the panorama will have meant much more than scenery to him.

When I reached the outcrop, I found a small cave with a wind-buffed entrance that was almost perfectly semi-circular, like a round-headed window. If Cuddy did indeed seek shelter in this shallow cave, he could not have done much more than sleep there. I crawled in and could barely sit up, although there was welcome relief from the breeze, even though the open entrance is wide. But it was only when I climbed to the top of the crag that I understood why this place would have attracted Cuthbert.

Jutting out from the hillside, commanding wide and long views, this was a place to pray and a place where God could see His servant searching the sky, arms aloft for His grace. And for Cuthbert, his intimate knowledge of the New Testament guided his actions and attitudes as he knelt on the rock, gazing out over the plain and the hills. Christ's Last Temptation took place on a mountain from where He could see all the nations of the world and where Satan tested Him with the promise of secular power. On his mountain, Cuthbert will have drawn strength from Jesus' example as he fought his demons.

About five miles to the west, at the foot of Yeavering Bell, lay Ad Gefrin, a Northumbrian royal palace and a focus of secular power in the second half of the seventh century, the sort of power Satan dangled before Christ. In its timber halls kings gave judgements, held councils and feasted with their warriors. Close by stood a remarkable structure that was probably used by royal officials, perhaps priests, perhaps the king himself. Known as the Grandstand, it was made from wooden beams and resembled a wedge cut from an amphitheatre. Tiered seating rose up in a triangle from a focal point on the ground, where orders were given by someone in authority. In a pre-literate society, it was vital that important people such as noblemen all heard the same thing at the same time.

On still days, Cuthbert will have seen the smoke of cooking fires rising from the royal palace at Yeavering and also from the rich farmland of the banks of the Till and its tributaries. On top of the crag, the winds of many millennia have carved curious wave-like grooves in the rock, but one of them might have looked providential. On the edge of the flat summit of the crag a small rainwater pool has formed, surely put there by the hand of Almighty God and filled with His holy rain so that Cuthbert could both drink and bless himself.

Following sheepwalks and the occasional track, I climbed above Cuddy's Cave and looked over to the wastes of Doddington and Horton Moors, a bleak landscape but a place that had once been touched by the gods. The map is speckled with what prehistorians call rock art, mostly cup and ring marks chiselled on outcrops of bedrock. Their precise function will never be understood, but the presence of at least nine sites of rock art, as well as a stone circle, enclosures and other signs of prehistoric settlement within less than a square mile means that this moorland was thought to be close to the gods long before Cuthbert walked there.

When shepherds saw him in the hills, they will have wondered at the identity of the wanderer before they could get close enough to see the tonsure that marked Cuthbert as a monk and not a fugitive or a madman. As he passed his days and nights in prayer, fasting and keeping vigil under the open, starry skies, living a life of solitude in the deserts of moss and tussock, it may be that farmers and herdsmen gave the holy man gifts of food, ewes' milk and perhaps sheepskins to ease his shivering sleep in the little cave and other rocky shelters. They will have known that he was praying for their salvation as well as his own, and interceding with God to send good weather and bountiful harvests. When Cuthbert later retreated to Inner Farne, both Bede and the Anonymous biographer detailed the provisions he made to feed himself, and even up on the high moors he will have considered how to survive. He had no land to grow anything, and it is likely that he depended on the charity of the countryside and its people. Perhaps they conferred the place-names that seem to remember his wanderings: Cuddy's Knowe, St Cuthbert's Grove, Cuddy's Well, Cuddy's Cave and others.

Attitudes to those who roam, live outdoors and shiver under the stars have fluctuated and my own ambivalent views were brought home to me at the beginning of the bitter winter of 2017–18 when I took my dog out for a walk early one morning. It was a luminous piece of orange plastic sheeting that alerted us to something that did not look right in the corner of the Haining Wood, where the gate opens into the Deer Park. When the sheet moved, Maidie started barking and bouncing. Then the shape of a man sitting behind the trunk of one of the stand of sweet poplars became clear. As we approached, he hailed me and said how much he liked the dog, even though it was growling at him. This greeting undammed a torrent of talk that moved between grand political conspiracy and the malign nature of society in general before swerving swiftly and seamlessly to the beneficial, health-giving properties of live yoghurt, especially when mixed with turmeric and Chinese spices.

Clearly well educated and equally clearly unstable, this man had slept the night in the woods, using the orange sheet to stay dry. Wrapped around his legs was a sleeping bag and he wore what looked like a new and clean padded navy jacket. He had pulled its hood up and was eating something from what looked like a cream cheese carton, the silver foil turned back. When I asked him (in the moments when he took a breath from ranting about the closure of a hostel in Edinburgh) which way he had come, he pointed to the dense woods behind him. Having shown him where the strands of the electric fencing ran, and not wanting him to hurt himself, I suggested he go back the way he had come.

I didn't think he was dangerous, except perhaps to himself, nor would injure or spook the animals, but I am no expert. He must lead a harsh and lonely life, and I felt sad that I was weighing him as a potential threat, suggesting he go back the way he came in part because I did not want him to see where we lived. If I had had any cash on me, I would have given it to him. He seemed older, perhaps in his fifties, although it was hard to tell with his hood hiding his hair. Hoods can make people look sinister. I called my neighbour at Middlestead Farm, Andy White, and he said he would check on him. I told Adam to lock all the doors. 'If tonight is cold or wet, the man might seek some shelter.' Later, I decided to go up and have a look for myself. I took some cash, but he had gone. There was no trace of him, not even an area where the grass had been flattened. I wondered what he did with his days. Perhaps the grim business of survival constantly occupied his thoughts and actions. In the snows, rains and icy winds of the bitter winter that followed, I sometimes thought about the wanderer who slept in the Haining Wood.

The night before I left to complete the last part of my journey with Cuthbert, I sat down with my maps and traced a route to the Kyloe Hills and St Cuthbert's Cave, a place where the monks fleeing from the Viking and Danish raids on Lindisfarne were said to have sheltered with his coffin and all of their relics. From there I planned to walk the St Cuthbert's Way to the coast and then cross the causeway to the Island of Tides. My calculation of the distance, the difficulty of the terrain and the time it would take to walk had to be precise because if I missed the safe crossing times, Lindisfarne would become an unreachable island and I would be stranded on the mainland.

My original intention had been to hire a campervan for a week, park it on Lindisfarne as somewhere I could sleep, write and, most importantly, be alone. But just as in 1965, camping of any sort was not permitted and so I was forced to book into a hotel. All of the self-catering accommodation seemed to be taken or far too expensive for just one person, but the hotel I chose had no bar or public area. That probably meant it would be quiet.

With enough clothing to last a week, maps, notebooks and other essential bits and pieces, my rucksack was full to bursting – and heavy. Given all of my tedious frailties, I wondered if my walking pace would be significantly slowed. I resolved to stop only when I wanted to take photographs with my phone.

From the maps, the village of Holburn looked like a good place to begin the last part of my journey. A sign attached to a blue door on a small, hut-like building encouraged me. It read:

Holburn Water Supply Trust. Built after a long debate and much wine by the cooperation of all the Holburn villagers. *Aqua Vincit Omnia. Rusticemur et Bibeamus*, 1996.

It made me smile. At our farm we have a private water supply that has recently been threatened by a developer, and discussions with our neighbours who share the system (councils of war, more like) have often been fuelled by wine. A bit too smugly I noted that '*bibeamus*' had been misspelled with an unnecessary vowel wandering into the first person plural. It means 'Let us drink'.

Curious as to why I had been curious about the sign, a villager who had been working in his garden approached. 'Are you walking to St Cuthbert's Cave?' His advice on the best and most scenic route turned out to be excellent and I walked past the dozen or so houses in Holburn on what was becoming a bright and warm morning. The main watershed ridge of the Kyloe Hills runs approximately south to north and beyond lies the coastal plain and Lindisfarne. As I walked closer I could see a series of shelved rocky outcrops with gaps and overhangs between them. One was pierced by a door-shaped cleft large enough to enter and it occurred to me that Cuthbert could have sheltered from the wind and rain in several places. The landward side of the ridge also backed into the prevailing wind.

Only wisps of cloud flitted across an open blue sky and after a mile or two of gently climbing I fished out my hat. A track led diagonally up the flank of the ridge and, as I neared the watershed, I turned to look out to the west. Even though I was not yet at the highest part of the Kyloe Hills, I could see even further than from Cuddy's Cave. Beyond Yeavering Bell, I was astonished to make out the three Eildon Hills shimmering, just discernible in the morning haze. As the crow flies, they must have been more than thirty-five miles away. And when I finally reached the top of the ridge, I looked east to the sea. Lit by the morning sun, low-lying and green in a blue sea, lay Lindisfarne. At last, journey's end, and perhaps a new beginning.

I could make out the castle rock, the village and the sand dunes to the north. Immediately breaking my self-imposed rule, I stopped to stare at what seemed like a mirage. It occurred to me that on this low hilltop Cuthbert could have seen Eildon Hill North watching over Old Melrose, the place he had sailed away from privately and secretly, and when he turned he beheld the island of Aidan and the monastery he founded. Amongst these craggy little hills and the moors to the west, I believe Cuthbert lived a hermetic life for a time, making his covenant with silence as he fought devils and sought to move closer to God. Perhaps in this liminal place, between the tumult of the world to the west – the palace at Yeavering and the ecclesiastical politics of Old Melrose – and to the east the holy island of Lindisfarne, he may have wished to pass all of his days. And, like Aidan, be borne up to heaven in the arms of angels.

As I climbed a little higher, the coastline opposite Lindisfarne came gradually into view and I could see that vehicles had begun to cross the causeway. At the concrete trig point, I put down my rucksack and pulled out a body warmer. It was a blessed relief, for I was beginning to feel the weight of it. But it led to other frustrations. Getting the straps correctly aligned so that I could hoist the pack back on my back was maddeningly difficult. They constantly hanked on my sleeves or my watch strap as I bent over, cursing and huffing. Below the trig point, a smaller outcrop of rock on the other side of a stone dyke had a cairn on it and I decided on a brief diversion. I like cairns. But before I could find a good stone to add to it, I came across the body of a kestrel. It had been brought down, I suspect, by a shotgun and then later eviscerated by some carrion-eater.

In the woods to the south of the Kyloe Hills, I climbed slowly and carefully down to the site of St Cuthbert's Cave. It is a spectacular freak of geology. High, long and about twelve or fourteen feet deep in places, the rock that over-hangs it looks as though it has been pulled out like a drawer. The vast weight of this roof is held up, it seems, by an impossibly slim column of much-eroded stone. If the fleeing monks did indeed stop here, many of them and the precious relics they carried would have found shelter in this capacious natural room.

On a large, rounded boulder by the entrance to the cave, I saw that a headstone-like inscription had been carved in memory of Mildred and Ernest Leather, the latter dying in 1916. What amazed me was that two further inscriptions had been carved to Vivien Leather in 1997 and Anne Berens (her married sister?) in 2001. Perhaps their ashes were scattered on the site. St Cuthbert's Cave is cared for by the National Trust and they would no doubt have frowned on its appropriation as a memorial to a family. And yet someone had arrived, presumably in broad daylight, with a mallet and a chisel – and left the name of the family responsible for a strange variety of vandalism. Very surprising and intriguing. Inside the cave, mostly on the back wall, there were a great many more names and some dates. Some were sets of graffiti-like initials daubed or scratched onto the rock, but others had been cut with a chisel, some skilfully. There were even several runic inscriptions. This leaving of names at a place associated with Cuthbert was something I was to see often.

Well signposted and with good going, St Cuthbert's Way led me down the eastern slopes of the Kyloe Hills past a large plantation of mostly evergreen forestry and towards the village of Fenwick. I reckoned that I had at least four hours to reach the causeway before the tide started to come in and I fell into a metronomic rhythm of walking quickly, trying to ignore the weight of my rucksack, the pain in my left shoulder, and my dodgy leg and back problems. Perhaps leathery old Drythelm would have approved of all of this discomfort, the twenty-first-century equivalent of a hair shirt. Good for my tattered soul.

At a place appropriately named Blawearie, St Cuthbert's Way became a C road that led pleasingly downhill to Fenwick. As I marched through the village, I noted that one of the houses was called Cuthbert's Rest. Not quite yet. At the foot of the main, in fact the only, street in Fenwick lay a very busy T-junction. The A1 thundered past and brought me abruptly back into the present, as articulated lorries hurtled north and south. I had to wait a long time to cross safely. A winding country road took me down to Fenwick Granary, where an old mill seemed to be in the process of renovation. I hoped they would install good soundproofing to screen out the near-constant drone of traffic. An old metalled track turned uphill off the C road and gave me a view of the seashore that told me I had allowed ample time. But I had another problem, something that was urgent. There being no public toilets and, I fervently hoped, no other walkers following me, I squatted down behind a bush and reflected on dignity and its departure.

Uninterrupted and unobserved, except by some well-fed ewes, I tightened my belt and did not slacken my pace. Others must have been more leisurely, for on top of a fence post someone had left an empty mug. The uphill path is marked on the Ordnance Survey as the Fishers' Back Road and is clearly old, with cobbled metalling that helped me keep up a good swinging rhythm. Once I reached the top of the path, Lindisfarne seemed comfortingly close, the coast only three wide fields away and the beginning of the causeway only a few yards further on. It seemed at that point I had wildly overestimated how long it would take me to walk from the Kyloe Hills and Holburn.

As I turned downhill along a field edge, the Edinburgh train suddenly hooted and hurtled north on the main line. I had forgotten how close to the island the railway runs, but

I could not yet make out how the path crossed it. At first I thought there must be a tunnel, but when I walked closer I saw that it was a pedestrian level-crossing, something I had never seen before. Unmissable signs told me I was at Fenham Hill, where trains travelling in excess of 100 miles an hour passed, and that before crossing I had to phone the signalman. There was a small yellow cabinet with a handset in it. Sadly but sensibly, there was also a sign put up by the Samaritans with a phone number to call. 'Talk To Us if things are getting to you'.

The signalman answered immediately and told me to stay off the line and call back. Moments later, the London train thundered past at a blistering speed, shifting the air so violently that I rocked back on my heels. When it had passed, in a matter of seconds, I called again and this time the signalman asked me how long it would take for me to cross the line.

'About twenty or thirty seconds,' I said.

'OK, you can cross now, if you go immediately and call me back when you reach the other side.'

Which I did. But very soon after I replaced the second handset another train travelling at high speed raced past. It seemed that escaping the tumult of the twenty-first century was going to involve drama as well as distance.

Once across the intervening fields, I came at last to the coast and began, it seemed, to pass through more barriers. Behind me I could still hear the hum of the A1 and another train as it rattled and clacked northwards, but the sounds of engines were slowly fading. At the seaward end of the last field, I sidestepped through a kissing gate to find myself walking between memories of warfare. Strung out in two long parallel lines were massive concrete cubes, part of the North Sea's coastal defences in the Second World War.

Designed to prevent tanks and other vehicles coming ashore and creating a mainland bridgehead for an invasion force that had sailed from German or Dutch ports, they looked bluntly immovable, an enduring monument to victory in a just war fought against the manifest evils of Nazism. Far from seeing the huge cubes as a disfigurement in a beautiful, natural landscape, I understood them as part of the means by which such beauty was preserved and made accessible to all. At the end of the rows of cubes, where the road from Beal cuts through to the beginning of the causeway, a large, red sign warned, 'No Shooting'. Not something the Wehrmacht would have paid much attention to as their Panzers rumbled into the heart of Northumberland.

Since the beginning of written record, war has lapped around the shores of Lindisfarne. Spreading out my unwieldy Explorer map on one of the concrete cubes, I wanted to find the outfall of a stream called the South Low (pronounced like 'how'). It was there that a pivotal episode in another war for Britain took place.

Ten years before the slaughter by the River Swale at Catterick in 600, a coalition of native British kings rode with their warbands to Lindisfarne, their ponies cantering on the hard sand of the beach, their pennants fluttering in the wind. Led by Urien Yrechwydd, the Lord of the Tides and King of Rheged, and with allied contingents commanded by Riderch Hen, King of Strathclyde, Gaullauc, King of Elmet and Morcant Bwlch, a prince who may have ruled on the Tweed, the British army had fought a successful campaign against the hosts of the Anglian invaders who had built a fortress on the rock at Bamburgh and begun to extend their reach over its hinterland. These ancestors of Cuthbert and the Northumbrian kings had originally sailed from Angeln in southern Denmark in such numbers that Bede wrote of their homeland being deserted. It may have suffered badly from flooding. Angeln means 'Hookland'. The first element of the name survives in angling, the more precise term for fishing with a hooked line. The Angles from Angeln gave England its name, but history might have produced an alternative rendering – Hookland.

For the doings of kings, saints, farmers and ordinary people of the sixth, seventh and eighth centuries, there is scant written record, with the sole really solid source being the work of Bede. His *Ecclesiastical History of the English People* was exactly that and it paid only occasional attention to the native British and their kings. There are long gaps of silence in what is rightly known as the Dark Ages, and so any and all sorts of texts have to be considered, some of them undoubtedly doubtful.

One of these rare sources is both infuriating and fascinating. At some point in the ninth century, probably in a remote monastery somewhere in the Celtic west of Britain, Nennius or Ninya sharpened his quills and compiled the *Historia Brittonum*, the 'History of the Britons'. Unblushing, the monk wrote in the preface that he had 'made a heap' of all the material he could find: genealogies, bardic sources, traditions, stories involving dragons and giants, and some reports of events that may have actually taken place. The latter mostly sound and feel plausible, and these may have formed part of a lost manuscript, a 'History of Northern Britain', that was probably written in Strathclyde, the kingdom of Riderch Hen. Had it survived, it may have complemented Bede's account of the English people. Here is an important passage:

> Hussa reigned 7 years. Four kings fought against him, Urien and Riderch hen, and Gaullauc and Morcant. Theodoric fought bravely against the famous Urien and

his sons. During that time, sometimes the enemy, sometimes our countrymen were victorious, and Urien blockaded them for 3 days on the island of Medcaut.

Hussa and Theodoric were successors of Ida, the first of the line of Northumbrian kings, and he had made Bamburgh his capital place, building a stockade on the great seamark rock. Its name changed from Din Guauroy when it became Bamburgh, originally Bebbanburh, and named after Ida's queen, Bebba. The passage from Nennius uses another old name that has the shadow of a story behind it. Old Welsh was spoken across Britain in various dialects, and over the four centuries of the province of Britannia it borrowed a great deal from Latin, often by simply adapting terms and names, disguising them with what the ignorant see as eccentric spelling, the use of consonants as vowels and so on. Ynys Medcaut is simply a rendering of Insula Medicata.

Urien appears to have been a dominant figure in the second half of the sixth century. Ruling over Rheged, he was able to create and lead a British coalition as well as prevail in at least two battles celebrated by the bards. The evidence for any sort of history of Urien's post-Roman kingdom in the north is gossamer-thin. Place-names offer hints of how far his writ ran. To the west of Galloway, near Stranraer, is the village of Dunragit and it derives from Dun Rheged, the Fort of Rheged. It lies close to Luce Bay, the sea and the wide waters of the Solway Firth. Along its northern shore are the remains of two fortresses at the Mote of Mark and Trusty's Hill that were probably garrisoned by Rheged's warriors. It was a sea kingdom communicating with itself by boat and trading far and wide with outsiders. Glass from the Rhineland and the Mediterranean has been found, as well as the crucibles and moulds used by jewellery makers.

Working mainly with bronze, which could shine to a deep lustre, they made brooches, belt buckles and other fine objects that could be worn. The coloured glass from Europe was used as inlays or for enamelling.

Near Gatehouse of Fleet, Trusty's Hill seems also to have been a royal stronghold of Rheged. In 2012, archaeologists uncovered evidence of more luxury objects and the remains of pottery containers that brought wine from central France. The southern shores of the Solway, what is now Cumbria, also seemed to form part of Rheged and it may have extended far to the south at one time. The ancient name of Rochdale in Lancashire is Recedham. While this last is a shaky assumption, the origins of Urien himself and the centre of his power will bear a little more weight. The name derives from Urbgen and it means 'born in the city'. And the city was Carlisle, the hinge of the sea kingdom of Rheged. Even when Cuthbert visited in 685, the fabric of the old Roman city had faded but had not been submerged in decay. Waga, the royal reeve or prefect, showed the saint a working fountain that must have been fed by an intact acqueduct and cistern. When a church dedicated to Cuthbert was built some time after 698, its alignment, still observable, sits not on the traditional east to west axis but follows the rectilinear street grid of the old city. The Roman walls still stood in the late seventh century and as late as the twelfth; the chronicler William of Malmesbury saw a large arched building with an inscription to Mars and Venus on it.

The glories of Rheged, the prowess of its warbands, the opulence of its jewellery, and the feasts and wine-drinking in its halls, were fleeting. Archaeologists working at both the Mote of Mark and Trusty's Hill have found clear evidence that they were destroyed by fire at the end of the seventh century, only a handful of years after Urien's army saddled their ponies and rode to Lindisfarne.

In another passage from the *Historia Brittonum*, Nennius noted that the British kings and their warbands made camp at Aber Lleu and my trusty edition of *Y Geiriadur Mawr*, the Great Welsh Dictionary, translates that as 'the Estuary of the Low'. Passing more rows of concrete cubes, I found the little stream easily enough as it flowed onto the sands by a rocky outcrop marked on the map as Beal Point. A necessary source of fresh water for men and horses, and rising ground that gave a good vantage point to watch for movement from the Angles blockaded on Lindisfarne, it was a good place to make camp. Urien and the British kings had defeated Theodoric's warbands, driven him to seek refuge on the island and had come to the Low to drive him into the sea. Here is another passage from Nennius, one that relates what happened next:

> But while he was on the expedition, Urien was assassinated, on the initiative of Morcant, from jealousy, because his military skill surpassed that of all the other generals.

The action and its motivation seem clear enough, but bardic sources add a little colour to the murder on the banks of the little stream. Urien was killed in his tent at dead of night by Lloflan Llaf Difo, which sounds like a conferred name since it means 'severing hand'. The great king's head was cut off and perhaps displayed to confirm his death. The British coalition quickly dissolved, the Angles escaped from their island prison, and the whole history of Britain turned decisively in a different direction. If Urien had lived and expelled the Angles from the north, England may have become known as Saxony and been a smaller province south of the River Trent, the capital of all Britain might have become York, and we might all be speaking another language.

After Urien's murder his son, Owain, succeeded and won victories against the Angles. But the tides of history were running against him and the retreat of native British power to its last western heartlands began. In response to sustained adversity, a surprising, heroic tradition grew in Wales and elsewhere. As the Angles and the Saxons in the south overran more and more territory, the bards never allowed their people to forget that Welsh-speaking kings once ruled in London. They called the English-controlled parts of England Lloegyr and it means 'the lost lands'. In Dyfed, during the reign of Hywel Dda, a poem known as the 'Armes Prydein Vawr' (the 'Prophecy of Great Britain') was composed. It called on the British of the Old North in Strathclyde, in Cumbria and in Cornwall to unite with the Vikings under the Banner of St David and drive the English back into the sea.

Alongside this, there grew a hope for a redeemer, Y Mab Darogan, the Son of Prophecy. He would emerge to lead the Welsh back to their ancient glories, to victories over the hated Sais, the invading English who stole the Holy Island of Britain from its rightful rulers. The bards sang of nine Sons of Prophecy: Hiriell; Cynan; Cadwaladr; Arthur; Owain ap Urien; Llywelyn Vawr, the medieval Prince of Wales; Owain Lawgoch; Owain Glyn Dwr; and Henry Tudor.

For almost 1,000 years, the Welsh yearned for their redeemer, a great leader who would march at the head of a host deep into Lloegyr. Ordinary people gathered on hillsides to hear tales of the heroes and the warbands of the past – of Cynan Meriadoc, the leader of the migration to Brittany, Little Britain; of Cadwaladr, King of Gwynedd; of Arthur and Llywelyn Vawr – and to hear prophecy. Owain of Rheged gained fame in Wales because the bards kept his name alive as one of Y Mebyon Darogan, leaders who would ride out to defeat the English.

The ebb and flow of events in the Dark Ages is rarely as clear-cut as the events of the seige of Ynys Medcaut. The picture is often very blurred, confusing, even contradictory. Apparently important names sometimes only appear once and nothing more is heard of them. Some academic historians pay attention only to Bede and other textual sources they think reliable and dismiss the bards, the genealogies and the traditions associated with native kings and their homelands. Nevertheless, twenty years of reading and writing about this fascinating period, the time when the nations of Britain were forming and its dominant languages emerging, have led me to some conclusions that are more than tentative.

The British – and Anglo-Saxon – kingdoms were not like modern states with defined borders, separate jurisdictions and different institutions. They were based on military power, on the royal warband, and loyalty. The more successful in battle or in raiding, especially cattle rustling, a king was, the more he would reward his warriors with gold, horses, privileges and other items of value, and his generosity encouraged more warriors to rally to his standard. The jewellery made on the Mote of Mark was not only for conspicuous decoration, worn like medals, it was also the currency of trust and obligation. Here is a passage from Taliesin, the great bard of the people the Welsh still call Y Gwyr Y Gogledd, the Men of the North:

> Urien of Yrechwydd, most liberal of Christian
>   men,
> Much do you give to men in this world,
> As you gather, so you dispense,
> Happy the Christian bards, so long as you live,
> Sovereign supreme, ruler all highest,

> The stranger's refuge, strong champion in battle,
> This the English know when they tell tales.
> Death was theirs, rage and grief are theirs,
> Burnt are their homes, bare are their bodies.

All that exists to characterise the culture of these post-Roman kingdoms are the praise poems of their bards, and of course that is a picture painted with the vivid colours of exaggeration and superlatives. And the everyday lives of the farmers, servants and slaves of Rheged, Strathclyde, Gododdin and the other native realms are not the stuff of poetry. Bards were fired by the deeds of warriors and in a passage from the epic poem composed to commemorate the battle at Catterick, brave Gwawrddur was in the van, fighting in the alder-palisade, the shield wall. This passage is also famous as one of the earliest mentions of the warlord Arthur, leading some (myself included) to believe that he was one of the Y Gwyr Y Gogledd, the Men of the North:

> He struck before the three hundred bravest,
> He would slay both middle and flank,
> He was suited to the forefront of a most
>   generous host,
> He would give gifts from a herd of horses in
>   winter,
> He would feed black ravens on the wall
> Of a fortress, though he were not Arthur,
> Among the strong ones in battle,
> In the van, an alder-palisade, was Gwawrddur.

Despite the tone and substance of these heroic verses, the small armies of the Angles and the British were not ethnically defined. The reference to strangers in Taliesin's poem may hint at that. Kings made alliances that benefitted them and their people and did not fight somehow to preserve Britain and its culture. Their motives were not often those of the myth-wrapped Arthur. Angles fought in British armies and the British sometimes made alliances with the Angles and Saxons if they deemed that to be to their advantage. The leader of the Gododdin coalition that was annihilated in the battle at Catterick was Yrfai, Lord of Edinburgh. His full name betrays his Anglian origins, for it is Yrfai map Golistan, a Welsh version of 'the son of Wulfstan'.

My work in ancestral DNA has added some statistical substance to the rapid transformation of society at this time, when history raced across the landscape. Numbers are notoriously difficult to estimate at a distance of fifteen centuries, but if the overwhelmingly male population of Anglo-Saxons in Britain was only 10 per cent of the total in 550, geneticists reckon that within only five generations it could have risen as high as 50 per cent. The Anglian and Saxon warriors who began to dominate in the east of England almost certainly took native wives and they quickly subjugated and outbred indigenous males. Simple reproduction and not genocidal slaughter could have made incomers dominant in less than a century, spreading their language and culture widely and quickly.

Nevertheless, there were some important distinctions between natives and incomers, particularly in the sixth and early seventh centuries. Not only did the British see themselves as Y Bedydd, 'the Baptised', fighting to preserve the light of God against the pagan darkness of Y Gynt, 'the Gentiles', the Angles, there was also an ancient political and cultural bond that occasionally united the British. Cymru, the Welsh word for Wales, and Cymry, the Welsh, both derive from a Latin term, *combrogi*. Loosely, it means 'fellow countrymen' and it harks back to the Roman Empire, the old province of Britannia and a sense of citizenship. Cumbria remained Welsh-speaking for centuries and its name also comes from *combrogi*. And this notion of an ancient community and its hoped-for restoration is embedded in the long traditions of the Sons of Prophecy.

As the Northumbrian kings grew more powerful in the first half of the seventh century, two elusive names suggest that there were close links with the fading prestige of Urien's kingdom of Rheged. Nennius noted something remarkable. It appears that King Edwin, the first Christian king of Northumbria, was baptised by Rhun, the monkish son of Urien, and brother of his successor, Owain. He may have been Bishop of Carlisle. Bede ignored this inconvenient link with the Celtic past, claiming that the king was converted by Paulinus, a missionary sent north from Canterbury. And most importantly a priest in the Roman tradition.

Rhun's brief appearance is a reminder of Rheged's Christian traditions. They were venerable long before Aidan founded the monastery on Lindisfarne. Some time around 450, a man called Latinus was grieving over the loss of his little daughter and he had a stone inscribed in her memory and set up at Whithorn. This grief-stricken father is the first Christian in Scotland whose name is known. At about the same time, Ninian founded the church at Whithorn, probably on a mission from the Christian community at Carlisle. It was called Candida Casa, the 'shining white house', probably because it was built in stone 'after the Roman manner'. Soon after that, another church remembered by a series of inscribed stones was set up at Kirkmadrine. The name is a homage to St Martin of Tours, a man Cuthbert and Bede admired and who is generally thought to be the pioneer of monasticism in the west. This network of Christian communities in

Rheged was literate and valued the fading memories of Rome as the Empire collapsed and Romulus Augustulus, the last emperor to reign in Italy, was finally deposed in 476.

As I walked back through the lines of concrete cubes, I could still make out the hum of the A1 behind me. But across the wide sands, Lindisfarne looked far, far away, much farther than a mile or two. Perhaps this was part of the process of it becoming a place apart. And maybe the sounds of the twenty-first century would be carried away by the winds. Before I reached the beginning of the causeway, I had to run a gauntlet of warnings and prohibitions. Not only was there to be no shooting, there would be no drowning either. A poster carried a photograph of a half-submerged car and loud letters said 'THIS COULD BE YOU. Please Consult the Tide Tables'. I had, and reckoned I had at least two hours before the sea reclaimed the sands. And so I set off, following in the footsteps of Cuthbert and of a fifteen-year-old boy I once knew.

It was mid-morning and traffic was pouring across the causeway: delivery vans from the supermarkets, an oil tanker, several vast tourist buses with blacked-out windows and very many cars. Hundreds of vehicles passed me and I had to get off the road often. Everyone seemed to be rushing, anxious to get off the causeway and on to the sandy *terra firma* of the island. This was not at all what I expected, especially in late September, the tail end of the tourist season. I had imagined an altogether more tranquil introduction.

But as I made progress, the sound of the A1 grew fainter and the seaweedy smell of the sea stronger. When I reached the refuge, I saw that its stilts and steep stairs stood next to a large pool of standing water. This was part of the course of the South Low. An inscription on the bridge across it surprised me. It was built in 1954. Before then, Lindisfarne was much more of an island and I discovered later that most goods and people moved between it and the mainland by horse and cart. Passengers in the two taxis that plied the crossing were advised to lift up their feet to keep them dry as they splashed across the sands and through the pools. Apparently the taxis did not last long, quickly rusting as seawater ate at their bodywork.

The traffic on the causeway did not slacken and I decided to take the direct route and follow the Pilgrim Poles. These were set up in the nineteenth century along the line of much earlier markers known as the monks' stones. The Poles run arrow-straight across the sands and make landfall at a place called Chare Ends. But when I set out for them after the bridge over the Low, I found that the sand was sinking more and more with each step. Perhaps I was going the wrong way because I could see two groups of people following the Poles far in the distance. I tried another approach but the sand was muddy and very sticky. Having only a limited time to get across, I decided upon discretion and returned to the tarmac of the causeway.

The sun had climbed high and once again I shed my body warmer. The shore breeze was perfect; cooling but not cold. I began to swing in a rhythm towards the tussocky grass of the island and the dunes beyond its fringes. When I finally reached dry land, I found that my memories from 1965 had not been accurate. On that frantic day I thought the road rose up a little as we splashed on to the end of the causeway but any incline was too slight for me to notice. But one thing had not changed. Unmissable in bold black on white, a large notice read 'No Camping on Holy Island in Tents, Caravans or Motor Vehicles'.

PART TWO

LINDISFARNE

The Holy Island
of Lindisfarne

Causeway

The Snook

The Dunesland Links

Chare Ends

St Combs Farm

village

Priory

The Heugh

St Cuthbert's Isle

The Castle

The Waggonway

Cove Haven

Sandham Bay

Emmanuel Head

8

On the Island of Tides

After a long morning's walking, I made the mistake of thinking that by making landfall on the island I had arrived at my destination. There was still a long way to go. The causeway across the sands is about a mile long but the stretch that curves around the foot of the high dunes to the north turned out to be at least two miles more. Lindisfarne is a saucepan shape, and before I reached the main area of the island I had the length of the handle to walk. My rucksack was beginning to cut into my shoulders (why did I need to bring books?) and for some blessed relief I carried it like a suitcase for a while. Each time I looked up, the houses in the village looked about the same size. Was Lindisfarne slowly drifting out to sea? Would I ever reach it and be able to put down my rucksack?

On the land causeway, there were pools of seawater left by the retreating tide and that exposed another gap in memory. In 1965, I had the strong impression of dry land. Later I discovered that this stretch of road had only been metalled in 1964 because the queen was coming to visit. Apparently she arrived by boat but all the dignitaries who would line up to shake hands needed a smooth surface rather than hard-packed sand for their limousines. Perhaps the tarmac was higher when we squelched along it in wet socks and boots. Perhaps we were so knackered we didn't care.

Help, or at least encouragement and distraction, arrived with two swallows. In what was clearly a passage of play from two young ones, they flew around me, often in perfect synchronicity, wheeling and doubling back on themselves at astonishing speed. Their flight reflexes took them to within an inch of the surface of the road or the sides of the dunes. Sheer exuberance. The little birds lifted me, and instead of thinking how weary I was I began to notice the deposits of the tide: tiny white crabs, mussels and the dead waves of driftwood lying on both sides of the road. I suddenly realised that the hum of the A1 had faded completely behind me, somewhere after I had crossed the course of the South Low. Fewer cars were passing, and coming in the opposite direction were the delivery vans, having stocked up the pubs, cafes and shops. I realised I was hungry, very hungry. Reasoning that I would be arriving at a place where I could easily get lunch, I had not packed the usual cheese sandwiches. I wished I had.

After what dragged on like an eternity, I rounded the last few yards of the land causeway to arrive at Chare Ends, where the land rises abruptly and the road leads to the main part of the island. There I saw something that seemed to form part of an emerging pattern. On a rubble-built stone plinth that in other places would have carried a welcome sign, it said 'No Parking'. It looked like an altar to prohibition. On my weary way along the causeway, I had seen other signs warning of 'Danger, Former Military Target Area', of plant attacks from the non-native Pirri-pirri burr (problematic for dogs), more 'No Camping, No Busking, No Vehicles Except Permit Holders' and several other cautions about safety. Had I arrived on the Island of No? Not a warm welcome or a good start. Looking back down the line of the Pilgrim Poles, I could see that several groups were making slow progress across the sands that were drying out, and in so doing had avoided this barrage of negativity. Maybe there were signs saying 'No Praying'.

When I passed the main car park, hundreds of vehicles glinted in the sun and the four or five main streets of the little village were crowded with processions of visitors. I found my hotel and, too early to check into my room, dumped my groaning rucksack, stopped groaning and wandered around in the crowds, thinking I could be almost anywhere, any tourist destination. If I was to discover the essence of the peace of Cuthbert, it would not be amongst this day-tripping throng, and yet Cuthbert and his cult were the reasons why visitors started to come to Lindisfarne all those centuries ago.

Appropriately and gratefully, I found Pilgrims Coffee House and joined the queue for an excellent latte and a substantial scone (not microwaved), almost Scottish in its size and Cornish with the amount of clotted cream. The interior being full, I sat out in the garden with about twenty others and, like them, immediately fell prey to a flock of ruthless, predatory sparrows. Feeding off scone and other sweetmeat crumbs, they were very bold indeed. One hopped up on the edge of my table, and instead of waiting for crumbs pecked at my scone even though the plate was at my elbow. Gerroff! And then when I was rummaging around in my pockets looking for a hankie, another two stuck their beaks in my scone. A genuine nuisance, thirty or forty of these little scavengers fed ceaselessly on what was left or dropped and were unabashed when shooed away by the wave of an arm. Cuthbert loved birds and I had a feeling he was laughing at our discomfiture.

I abandoned my scone, finished the excellent coffee and went off to the post office to buy a newspaper, something I could read over lunch in a sparrow-free zone. Both choices were a mistake. Thinking that local is freshest and best, I ordered what was advertised as a Holy Island crab sandwich with ciabatta bread. The fact that only one other table in the restaurant was occupied and the sandwich cost £10 should have sounded alarm bells, for when it finally came the whole thing was inedible. Hard as a brick, the ciabatta would have broken a tooth, and the crab meat was hot, none of it white, while the garnish was a scatter of soft tortilla chips and a none-too-well-washed salad with a dribble of what might have been dressing. Or might not. Oh dear. No sparrows, but instead a predatory, touristy rip-off. After my long walk, the sofa was very comfortable and I pushed my plate away, sat back and opened up the newspaper. Another mistake. On this beautiful island, a timeless place of spirits and saints, I found I simply did not want to read about the Labour Party Conference or indeed about any of the other cares of the world that lay beyond the causeway.

When I left the terrible lunch, still hungry, and stepped back out onto the village streets, they were almost deserted. The tide. Of course. That day the latest safe crossing time was 13.50 and drivers were advised to set off well before then, since it might take longer to cross if cars and buses all departed at around the same time.

As I looked for a bin for my unread newspaper, I passed only four or five people and they seemed to be going about their daily business. As the line of cars streamed along the land causeway, the island seemed to let out a long breath and become itself again. Later in the week, I spoke to a lady at St Coombs Farm, not far from the village and the only farm on Lindisfarne, and she told me that without checking the tide tables or looking at a clock or watch, she knew when the tide shut. Something in the air shifted. But shut? That was the verb used by all of the islanders I met. The tide shut off the mainland, like shutting a door. Over my time on Lindisfarne, I realised that there were two islands and two histories to understand.

Since there would be no queues or crowds, I decided to visit the priory. I was too tired to walk far. Armed with a Mars Bar, I tried to buy a ticket. But in another repeating pattern, I found that the ruins of the medieval priory were closed to visitors when the tide shut. It was not because of a lack of demand. There were still people on the island who were staying overnight, like me. The ticket office and the visitor centre's shop and exhibition closed because the people who worked there were not islanders and had to cross the causeway with the departing day-trippers to get home. When I grumbled about this, an officious English Heritage official pointed out the small print at the bottom of the board that advertised a closing time of 5 p.m. – except in exceptional circumstances (that were entirely predictable) or some such rubric.

Beyond the visitor centre, a close led to the gate to the priory grounds and because it also led to St Mary's, the parish church of Lindisfarne, it was open. I passed boards that detailed with great formality and precision 'The Services Rendered by the Holy Island Life-Boats'. Dating from the early nineteenth century to 1965, it listed all of the notable call-outs. An impressive and moving roll of honour, with the date, the boat or ship that was rescued or assisted, and the number of lives saved: from the SS *Coryton* out of Cardiff, twenty-seven people were rescued on 16 February 1941, and on 27 August 1953, the yacht *Mermaid of Poole* was saved, along with three passengers. As I read this recital of bravery, selflessness and duty, I was not only awed but reminded that Lindisfarne was once a community that lived off the sea – its bounty and traffic – and that, around its singular beauty, savage seas could claim lives.

I passed the headstones in the churchyard, some of them recent, others bearing the names of several generations. I came across a very new stone, not at all weathered, small and with a gold star chiselled at its top. It read:

IN LOVING MEMORY OF
HUGO JOHN ROBERT GLENTON
20TH APRIL, 2017 – 30TH JUNE 2018
AGED 1 YEAR AND 2 MONTHS
OUR BEAUTIFUL BABY, FULL OF SMILES.
ADORED BY ALL WHO CUDDLED HIM.

And on the plinth:

WE TREASURE YOUR MEMORY,
AND THE MIGHT HAVE BEENS,
YOU'LL FOREVER BE WITH US
WHEN WE'RE LOST IN OUR DREAMS.

Jolted at the rawness of recent tragedy, I stood and stared at the little headstone, and after a time began to weep for all of the pain the parents of the wee boy suffered, perhaps the pain he suffered, the grief at the unnatural injustice of the death of a child before his parents, the rage at the loss of his future, and the dignity and power of what they had asked the mason to engrave in his memory. I hoped it was a shred of comfort in all that darkness to know that their baby was buried in ground where saints walked and prayed, in holy ground.

St Mary's Church is old. Some of its masonry dates to the twelfth century and the rest to the thirteenth. What still surprises me is its spatial relationship with the medieval priory, built at almost the same time. The two churches are no more than thirty yards from each other and are aligned end to end. My research had told me that this apparently awkward arrangement was characteristic of Anglo-Saxon monasteries and I could understand that both St Mary's and the priory church would wish to sit on the traditional east–west axis. Between them I saw a massive cross socket, much larger than the one I came across on the road to Dryburgh Abbey. So massive that it is unlikely ever to have been moved, it is a rare relic that survives *in situ* from the old monastery founded by Aidan in 635, the place where Cuthbert had been buried. And that probably meant I was standing close to its centre, a tall cross that was a focus of prayer, worship and ceremony.

Crosses like the one that fitted into the massive socket were not the plain, symbolic sort erected in modern church-yards. The early ones told stories. They were used as texts and illustrations by preachers. Engraved, often on all faces of the shaft, the crosses made by the Anglian masons of the kingdom of Northumbria were glorious, elegant and mystical objects of great reverence. Many years ago, I diverted from a journey to Stranraer and the Irish ferries to go and look at the Ruthwell Cross. Erected in the early eighth century, when Northumbria had absorbed the western kingdom of Rheged, it is displayed in a side chapel of a small church in a village between Annan and Dumfries. The gallery arranged around the cross allows visitors to get close and enjoy the virtuosity of the carving. More than sixteen feet tall, it has been recon-structed virtually entire and on the front and back faces of the shaft there are scenes from the life of Christ and episodes from the stories of the Desert Fathers. One of them may be St Anthony of Egypt, a man much admired by the Lindisfarne monks. The sides are decorated with interleaving foliage known as inhabited vine-scroll. Amongst the inhabitants are birds and what might be lions. All of the vignettes and biblical scenes, as well as the decorations on the cross-piece itself, would have been painted in bright colours: ochre reds and yellows, blues and greens from plant extracts. And standing outside in all weathers as these crosses did, the painting will have had to be regularly renewed.

Most intriguing, and what adds metaphor as well as mystery, is the role of the Ruthwell Cross in the early history of English literature. Written in Northumbrian runes cut on the lower, narrow sides of the shaft are two extracts from one of the earliest surviving poems composed in English, 'The Dream of the Rood'. It forms part of a highly sophisticated scheme, in both poetry and sculpture, for reading the whole cross as a kind of text. The poem is based on the notion that the cross on which Christ was crucified had a personality. Here is the second stanza:

I [lifted up] a powerful king –
The Lord of Heaven I dared not tilt.
Men insulted both of us together;
I was drenched with blood poured from the
    man's side.

The ancient beauty and power of the Ruthwell Cross prompted its destruction: it was smashed in 1642 by hard-line Presbyterian iconoclasts, men who found its imagery offensive because it smacked of popish idolatry. In 1823, the parish minister, Henry Duncan, picked up the pieces that had lain in the grass of the churchyard for almost three centuries, restored them and re-erected the old cross in the manse garden.

I left the grave of the baby boy, wishing I believed in prayer and its ability to comfort. All I could think to do was wish Hugo's parents well and hope that like my son and his wife they would be or were blessed with another child. A brother or a sister would be a joy in themselves but also a hand that could reach back into the darkness of the wee one's death to draw him closer.

Raised in the traditions of the plain, sombre kirks of Scottish Presbyterianism, I find churches to be depressing places that reek of disapproval and prohibition, and dwell too much on failure and death. To my youthful ear, hymns were mostly joyless dirges and sermons reminders of what a sinful, scruffy and generally unworthy boy I was. In Kelso we had an evangelical believer in hellfire for a minister. Thinking back on interminable Sunday morning services, they were very theatrical in the way they were set up. Perhaps my memory is faulty, and selective, since my attendance quickly moved from occasional to never, but the big moment in our kirk seemed to be the appearance of the minister on stage in his costume of jet-black flowing robes and stiff white collar. Maybe to the accompaniment of something uplifting on the organ (played by his tiny, bespectacled, stick-thin wife) or a sonorous hymn, he would slowly climb the stair to the pulpit, clutching a huge Bible to his chest, his face upturned, chin jutting and a mane of fair hair swept back off his fore-head. I prayed, perhaps the only time, for him to trip. The minister reminded me of Finlay Currie, a Scottish actor who played God in several Hollywood epics, but was not nearly so avuncular and benign. When I sat in the Roxy or the Playhouse riveted at the technological magnificence of these overblown films with their soaring scores, I liked that the Almighty had a Scottish accent and that the Romans all seemed to be Americans. I have a memory of John Wayne playing the soldier in attendance at the crucifixion who recognised that 'Truly, this man was the son of God', pronounced Gad.

When our minister reached the pulpit, he slammed the big Bible down on the lectern, opened it at the text for the sermon, and before he began he looked out over the congregation for a few moments, searching for outward signs of sin and noting absentees. And then it began, hellfire, burning pits, endless torment and all. As he worked himself into a righteous passion, the minister's face, florid at the best of times, began to colour even more as he leaned out over the pulpit to despair of our salvation. When it came to the blessing at the end of the service, I felt we all needed it.

I have been in English churches several times and know that they are generally much more cheery. But nothing prepared me for the warmth and wonderfully colourful, decorated and detailed interior of St Mary's, Lindisfarne. It was so welcoming, and a good place to be after having stood at the wee boy's grave. At the east end of the nave stood a stone font covered with fruit and vegetables, presumably put there for the Harvest Thanksgiving services. Apples, marrows, turnips, bananas, jars of jam, onions, tins of soup, flowers, a jar of curry paste, a packet of flour and all sorts of surprising items were piled on the top and on the base. Around the whole church there was fruit, apples lined up on the altars, and lovely displays of flowers.

On either side of the nave run rows of polished pews, well cared for, and stained-glass windows fill the church with rays of brightly coloured light. Most striking were two carpets, one laid over the steps up to the Fishermen's Altar in a side chapel and another at the main altar at the east end. Their designs were lifted from the *Lindisfarne Gospels*, from what are known as the carpet pages. Gorgeously coloured and decorated, these are abstract variations on crosses, very geometric and reminiscent of non-figurative Islamic art. According to Bede, some priests in Northumbria used prayer mats and these might have been actual oriental rugs or copies of their designs. The carpet in front of the Fishermen's Altar is a huge version of the carpet page from the St Luke's Gospel.

In high contrast is an impressive sculpture that stands in the south side of the nave. Carved from massive blocks of wood, six monks carry the coffin of St Cuthbert. It commemorates the period after 875 when the monastery is said to have been abandoned at the coming of the much-feared Great Heathen Army, a coalition of Danish and Norse warriors who had invaded England. The monks carried their relics, the coffin of the great saint and possibly even the timbers of Aidan's original church with them. The sculpture has real power, a study of faith and gritty determination, but its sombre note seemed to me to be discordant in that bright church. On the opposite wall of the nave are reproductions of the carvings made on Cuthbert's coffin on Lindisfarne soon after 698. They show Christ, the symbols of the gospel writers, Matthew, Mark, Luke and John, a very early representation of the Virgin Mary and Child, the Twelve Apostles and seven archangels. These were all labelled in a curious mixture of Latin script and Northumbrian runes, like the Ruthwell Cross.

The late seventh-century coffin was a work of great art that showed how Cuthbert and his contemporaries imagined their God, angels, the apostles and God's mother. Like the crosses, it was almost certainly painted and made even more vivid, even more of a rich focus for devotion. The massive wooden sculpture has power, without doubt, but in my view paint would have enhanced that – because it was clearly important to the monks of the seventh and eighth centuries.

Two competing perceptions seem to be at work here. We live in a world of high colour, seen every day in all sorts of ways: our houses are painted; so are the objects we buy; the clothes we wear are dyed; we see bright colours in publications, on advertising hoardings, on TV, almost everywhere we turn. In the early medieval period, the world looked very different. Since everything was made from natural materials, buildings, clothes and everyday objects all blended with the colours of the landscape, the tints of spring green, grey skies, autumn yellow and brown, and the blue and green of the sea. Often the only distant sign of a small settlement of wooden houses roofed with rushes were their spirals of smoke from their cooking fires. Vivid colour was rare and usually small in scale, like ripe berries and wildflowers, and I believe that this was part of the reason the gorgeously painted *Lindisfarne Gospels* were made, the crosses decorated and the images of Cuthbert's coffin picked out in rich reds, yellows, greens and whatever pigments were thought suitable or were available. Colour made these sacred objects even more distinctive, perhaps even more sacred.

Partly because surviving fragments of classical temples and their sculpture have come down to us in the neutral colours of stone, we forget that the likes of the Parthenon Frieze was painted and made to look as naturalistic as possible. The sombre look of classical sculpture seems to many in the modern era to be part of their dignity and beauty. But the Greeks, Romans and early medieval churchmen saw things differently. Paint did not look gaudy to them but was a celebration, another way of praising God or the gods. The monochrome wooden sculpture in St Mary's seemed to me to be a very modern way of honouring the piety of Lindisfarne's monks and their great saint. I think it would have surprised them.

When I walked over to the Fishermen's Altar to look more closely at the carpet made in homage to the carpet page for

St Luke's Gospel, I noticed another contrast with the austere church of my youth, but something Cuthbert would have recognised and warmed to. Below a reproduction of an early medieval Madonna and Child, there was a black wrought-iron table where rows of small candles flickered. Below was a packet of more tea-lights and a box for donations. Under the image of the mother and her baby, I lit candles for Hannah, Grace and Hugo.

Under a darkening sky and in a breeze that had freshened into a high wind blowing out of the south-west, I walked down to the shore below St Mary's. The incoming tide had surged over the sands quickly and spindrift blew off the tops of the whitecaps as the sea grew choppy. Below the southern edge of the priory ruins rises the Heugh, a long, flat-topped rock that shelters the old church and monastic buildings, and it is also a vantage point looking out over the waves and the Northumberland coastline. To reach the summit of the Heugh from the western end, there is a steep, rocky staircase put in place by the convulsions of the earth millions of years ago and made accessible by centuries of use. At first it looks like an awkward climb, especially when a strong wind threatens those who stray too near the edge of the cliff, but soon I realised that hundreds of thousands of pairs of feet would show me the safe way. Certain flatter rocks, spaced at the interval of a single step, have been darkened and made smoother by the soles of many, many boots, shoes and sandals.

The grassy summit of the Heugh is dominated by a former coastguard station with a covered observation deck at the top. By its side stand the ruins of the Lantern Chapel, an enigmatic building that may have had a maritime function, either as a seamark or as a vantage point, or both. Beyond it is a stark, metal-framed mast that is certainly a modern navigational aid used by those wishing to make safe passage into Lindisfarne harbour between the sandbars and rocks. Once the mast is lined up with the tower of St Mary's Church, then boats should be able to make landfall without mishap. Between the mast and the coastguard station stands the tall cross of a war memorial. But none of these upstanding structures interested me as much as what lay only a few inches under the grass.

In the summers of 2016 and 2017, archaeologists made dramatic discoveries here. Far from being a rocky whinstone ridge that merely sheltered the monastery founded by Aidan in 635, the Heugh may have been its spiritual focus. Like a vast natural altar overlooking the land and sea, and reaching up to the skies and heaven beyond, it was God's beacon, a place where a light might shine over the vastness of Creation. Near the war memorial, excavators uncovered the foundations of substantial walls that strongly suggest a high tower, perhaps as high as the modern coastguard station. According to Bede's moving account, the monks on Lindisfarne saw a beacon lit on the island of Inner Farne in the late winter darkness of 20 March 687. It was a signal that Cuthbert had died. Perhaps an answering light was kindled from the Heugh.

A year after the first dig, archaeologists opened a second site near the metal-framed mast and, once the soil had been carefully trowelled away, the outline of a small church was revealed. The nature of its construction and the lack of binding mortar strongly suggested that it had been built in the late seventh or early eighth centuries. And it may well be that this stone church was raised on the site of Aidan's original wooden building. Later documents and traditions spoke of two named churches on early medieval Lindisfarne. St Cuthbert's of the Sea was a chapel on Hobthrush, a small tidal islet not far offshore from the Heugh and the refuge where Cuthbert began to hanker once more after the peace of a hermetic life. In 2017, it seemed that archaeologists had discovered the second named church, which was known as St Cuthbert's of the Sky.

If Aidan chose this high, singular and apparently inhospitable places of winds and storms as the spiritual focus of the community he founded, then his reasons were not perverse, or difficult to understand. In the tradition of the diseartan, places apart from the world, the Heugh seemed doubly blessed, a gift from God. Already cut off each day from the tumult of the world because Lindisfarne is a tidal island, the high rock was further set apart because three sides of it are steep and it is easily accessible only from the east, and there through a narrow path. But more than all of that, it was a place on the edge of yonder, a place where God could see clearly those who prayed to Him and who praised Him. Cleansed by the sea-spray and purified by the incessant wind, this elemental, even harsh, place was a rock of ages but not one where the faithful would hide themselves. Fourteen centuries later a child understood what Aidan had understood. A prayer was recently left at St Mary's Church that read 'I love you God and Jesus, can you see me?'

And the rock is cleft. The sense of the Heugh as a natural altar is enhanced by the tradition of the Prayer Holes. On its seaward side, the steep cliffs of whinstone are indented by tiny, shallow caves, little more than niches where a man might press his back against the cold stone, kneel, look out to sea and search the sky for his God. The rhythms of the psalmody, prayer and meditation might have chimed with the shushing of the waves fifty feet below and the soughing of the eternal wind. And on all but the warmest summer days, the flesh will have been mortified and cramped by the cold stone, the spray, the rain and the whip of the breeze and the stinging sand it carried. When night vigils were kept in the Prayer Holes, the pious will have felt themselves on the edge of the world, between it and eternity.

When his brother monks first brought Cuthbert's body from Inner Farne to Lindisfarne, they will have been able to approach the Heugh and the monastery below it more closely. In the seventh and eighth centuries, the tides washed further inland than they do now because the waters of the harbour reached much nearer to the ruins of the priory. The eastern end of the Heugh was bounded by the sea on three sides. If Aidan's church was built on its summit and not, as has long been assumed, where the church of the medieval priory now stands, then Cuthbert's body will have been placed there, in the most sacred part of the monastery. And that was why I wanted to climb up to the Heugh after the tide shut, to be where he had been, in life and in death, to see what he had seen, to try to discover something of how they saw this place. Then perhaps I could begin to move closer to understanding the peace of Cuthbert.

Looking down at the ruins of the priory, the red sandstone warming in the late afternoon sun, beyond them to the village, the fields of St Coombs Farm and, on the northern horizon, the grassy humps of the sand dunes, I began to see what Aidan had seen. If the Heugh was an altar made by God, then the whole island was His church. Walked by saints, studded with painted crosses, sanctified by prayer and self-sacrifice, Lindisfarne was a vast sacred precinct. God and not King Oswald of Northumbria had summoned Aidan to His island, the saint had understood His purpose, and His church was built not where the ground was easy but up on the windblown altar where the waves of the world broke on the hard rock.

The sites of the summer digs had all been sensibly covered over and protected by the preserved sods of grass, which had been carefully laid back on top of the stone foundations. But I could clearly see where they had been. Close to the coastguard station, a very simple and beautifully cut stone bench had been built against the eastern wall of the Lantern Chapel and beside it stood an upright stone that carried an inscription, 'Wild Lindisfarne'. At this place of Cuthbert's end in 687, it reminded me strongly of the taller of the Brothers' Stones, the place where his journey had begun more than thirty years before.

I decided that it would be good to return to the stone bench on the Heugh after darkness fell and the island slept under the blanket of the night. Perhaps the silence would speak to me; perhaps in this place where the veil between worlds, between the living and the dead, is thin, I would hear history whispering. And I would return there in the grey light of the morning before the island awoke.

Growing cold and stiff after my day of walking, I decided on a short, early evening perambulation, one with plenty of places to sit, a journey around what I reckoned were the favoured routes for those with little time on the island. Now that the tide had shut, these might be more accessible, more peaceful. On my way down to the western end of the Heugh and the stone stairs, I had begun to notice inscriptions, not only brass plates attached to the backrests of benches, something often seen in municipal parks, but also carvings and plates on gates, on walls and at entrances of other sorts.

On a wooden field gate next to the priory walls, I saw something carved on the top rail in a font I recognised. I expected to read a biblical quote or something from Bede. But instead it was a homily from Robert Louis Stevenson, a truly great writer whose novels I devoured as a teenager: 'Don't judge each day by the harvest you reap, but by the seeds that you plant'.

Most of the inscriptions I read were simpler and some very touching. On another gate close by, Peggy was commemorated as 'a Friend and a Godmother', the brevity eloquent. Along the road from the village to Lindisfarne Castle, probably the most popular perambulation for day-trippers, many benches line the wall by the road and most are variations on an insistent refrain: 'Remembering Pauline Mary Cunningham of Nottingham, 1942–2009. She Loved This Island'. Again and again, those who donated benches so that others could sit down and gaze quietly at the view repeated similar sentiments, an enduring love for Lindisfarne. One recalled good times of a louder sort:

> In Memory of Banjo Bill Nelson, much loved Dad and Grandad
> 'We'll be going round the Island when we're done,
> 'And the merry Island rovers, Will all have big hangovers,
> But didn't we have a lot of fun.'

Another commemorative plate recorded a piece of recent history. On a bench close to the harbour:

> In great gratitude for the good welcome our five brave freedom fighters, Tormod Abrahamsen, Nils Havre, Sven Moe, Jan Stumph and Kai Thorsen, received here on Holy Island on 5th November 1941, after crossing the North Sea in a small boat. Presented on behalf of families and friends in Norway, 31st August 2010.

After the German invasion of Norway on 9 April 1940, more than three thousand very brave men and women crossed the North Sea to join British forces or become involved in clandestine operations in their occupied homeland. Tormod

Abrahamsen and his four friends pushed their twenty-foot rowing boat into the sea at dead of night near Kristiansand on the southernmost tip of Norway. In the dangerous waters of the Skagerrak, the heavily patrolled straits between occupied Denmark and Norway, they moved westwards, rowing without lights and probably without hoisting a sail, something that on the horizon might have given away their position. After three days in heavy winter seas, they made landfall on Lindisfarne. It was almost certainly a happy accident that saw their boat rasp up onto the shingle of the island's shore. Britain is a very long target and with only a compass to guide them, they knew that if they held a steady westerly course from the mouth of the Skagerrak, they would not miss.

The last bench I saw was the most affecting. Looking out over the harbour, where the crab and lobster fishing boats bobbed at anchor, I sat down to watch the shadows lengthen and the day wind down to its close. I felt my boots brush against something tucked under the bench and, creaking a little, I bent down to see that three bunches of artificial flowers had been attached to the lower struts with cable ties. Each had a laminated sheet attached with lines of verse on them, the sort of thing found on cards: 'They say that time's a healer, But that just can't be true, For no amount of time could heal, The pain of losing you.' Perhaps these were tied to the benches by people who could not afford to donate one, or had been told that there was no room to accommodate others.

All of these inscriptions and the many others I saw seemed to be modern variations on the ancient wish to be buried in sacred ground as a means of having the sins of the flesh washed away and making the path to heaven less arduous. In life, most of these people came to Lindisfarne because they saw it as a haven, a place to run to when life became difficult, a place to grieve, or find peace or simply joy in gazing quietly over the glories of the sea, the sky and the land. The dead felt close on Lindisfarne, and as darkness fell I sensed their spirits in the air.

When I called to organise my week on the island, my cheery hotelier had advised me to book somewhere else for supper since he only did breakfast, and apparently all of the hotels and other places to stay were full. There were three choices, reduced to two after a terrible lunch, and I had reserved a table for one at the Ship Inn. Earlier, I had bought a souvenir, a bottle of Holy Island gin, 'The Spirit of High Tide', and noted that it was an exciting 44% proof. The day's exertions persuaded me to open it when I got back to my room, but I had left it too late to buy a lemon and I had no ice or tonic. Walking up the Marygate, I realised that I would pass the Ship Inn and a large sign told me that they were also the distillers of the powerful gin waiting for me. They kindly supplied me with ice, a bottle of tonic and the chef insisted that it would taste better with orange peel. He went off to the kitchen to fetch some and the barman waved away any attempt to pay. All very cheering and a sharp contrast with the rip-off lunch. The chef was right, and since they gave me enough ice and orange peel for two glasses, it seemed a shame to waste them.

One of my earliest memories is of my dad coming home from work. The ritual seemed never to vary. In the back lobby I could hear him taking off his overalls and hanging them up in the coal cupboard before coming through to the kitchenette. My mum filled a basin with hot water and, stripping down to his white singlet, he washed his face, hands and forearms with great vigour, snorting through his nose and screwing his eyes shut against the sting of the green

Fairy soap bar. As he dried off, my mum placed a heaped plate on the table and handed him a copy of *The Scotsman*. In those days it was a broadsheet, and once he had folded it small enough he propped it against the bottle of HP Sauce. This was what my mum called a 'Reading Tea'. No one spoke as the food disappeared.

Being alone on Lindisfarne and wishing to avoid conversation, I knew that all of mine would be Reading Teas, and so on my way through the darkened streets to the welcoming Ship Inn I'd brought my novel (Carlos Ruiz Zafón, *The Angel's Game*), my maps and my notebook. In a warm, softly lit dining room, surrounded by the chatter of other diners (it was full) and the bustle of waiters, the dishing out and consultation of menus and the clatter of plates, I felt as though I sat in a bubble of silence, and to my surprise I liked it. I do like to work alone, partly because I don't like to be told what to do or be constrained or distracted by others, but I did think I would miss conversation in the evening. But I didn't. In fact, once I had finished an excellent dish of fried seafood, I asked for the bill so that I could get back outside into the quiet of the night.

As I walked down the Marygate towards the harbour and the eastern end of the Heugh, I realised that the only sounds I could hear above the silence were the wash of the waves and the metallic clink of a wire halyard against the mast of one of the fishing boats lying at anchor. Back home on our little farm, the silence sounds different as the wind rustles the trees, the old house creaks and a horse whinnies in its box.

Beyond the village, there are few street lights on Lindisfarne and the cloudy night made it difficult to see exactly where I was walking. So that I could climb up to the Heugh by the easy path at the eastern end, I had first to pass the harbour, and remembering my Chaplinesque ability to fall over or fall into things I used the torch on my mobile phone to light the way. Up on the Heugh the silhouette of the coastguard station was just discernible against the dark sky, and when I found the stone bench I clicked off the phone, pulled my jacket collar up and sat down. Having made my own covenant with silence, what voices might I hear?

I had read part of an interview given in 1987 by a man who called himself Brother Harold. He had led a hermetic life and saw it as a battlefield of constant conflict, just as Cuthbert did. Most important, he had learned not to turn away from God when life seemed to overwhelm him and instead used silence to come to terms with what he called the noise inside his head. After a struggle, he gradually replaced it with the word of God. Reaching back across thirteen centuries to Cuthbert and his battles with demons, the links and the language are striking.

As I sat up in the winds swirling around the Heugh, looking over the vast darkness of the sea, all I hoped for was to stop or at least quieten the noise in my head. I began to go over what I saw as the turning points of my life, hoping to discern some direction, perhaps even some purpose. The problem with that was my tendency to self-criticism, to focus on what I had failed to achieve rather than the good things that had happened. And yet that was part of the point of coming to Lindisfarne, my journey a mixture of pilgrimage and retreat. The island is a startlingly entire thing, a place apart which might offer me a detached vantage point from where I could look back over the sands of my life.

When I was thirteen, I recall what I would now call an epiphany. Then, it was known as catching a grip. At primary school, I had been a disruptive child, unwilling to listen or learn or behave, and I found myself in the handcrafts class, making things like raffia mats – in between riots and fights.

By contrast, my older sister, Barbara, was academically outstanding and eventually became Dux of the High School. A Latin title, the Dux medal is awarded to the brightest pupil in Scottish schools, and it is usually a reward for all-round abilities, sporting and performing as well as academic. My mum eventually felt compelled to do a brave thing for the times when all of her maternal instincts drove her to seek a meeting with the headmaster. She told him (I was in his study, instructed not to look gormless) that since my sister was clever, I could not be as stupid as I seemed, and it was the school's job to get me to use my brains – to educate me, in a very literal sense. To my amazement, and horror, the headmaster agreed and I was immediately promoted to the A class and told in words of very few syllables by my mum to bloody well behave.

I did, but only enough to survive and stay out of trouble. Very little homework was done, lots of lies told, and in those days of streaming and class rankings I was decidedly middle of the table, fifteenth out of thirty. But one evening, on my way to play or plot mischief with my friends from the old handcrafts class, I stopped on a path that led between the back gardens of our council estate. I can remember exactly where, and I gave myself a real talking-to, something about me deciding how my life would go, nobody else. I was not going to fail, be mediocre or be lazy. I was going to work, and I turned around, went home, opened my schoolbag and did my homework.

Up on the Heugh, I took myself back to 1963 and that turning moment, asking myself if it really was as dramatic as it seemed in retrospect. I suspected it must have been. My exam results began to improve, I found an aptitude for languages I have never lost, I started to compete rather than mess about at sport, especially rugby. I played as a schoolboy against Wales and found myself at Murrayfield in the Kelso team at only seventeen. I still have the cutting of Norman Mair's piece in *The Scotsman* that said I was the pick of the Kelso pack. And, not quite emulating my big sister, I was runner-up for the Dux medal at the High School. I remember my mum wanting me to 'get on' and 'stick in' so that I could live a better and bigger life than my parents. And I did that, but not only for them. I wanted it too.

In the silence of the darkness on the Heugh, I realised that I had lost focus on the course of my whole life and too often concentrated on incidents, usually things that had gone wrong, times when I had made mistakes or been let down by disloyal people, going over and over them. That backward view had to change because I could not change the past, and perhaps in my week on the island I might make a beginning, see how I might pull back from seeing my life as a series of problems, difficulties with enemies I had made or people I had hurt.

The wind began to whistle around the stone bench I was resting on and the wall of the ruined Lantern Chapel provided little shelter. Even though I was wearing four layers, including a warm woollen jumper and a good anorak, I began to feel the cold seeping into my bones. The eastern path leads to an old-fashioned cast-iron stile that gives access to a field next to the priory walls and, with my phone torch, I found it easily enough.

Walking back through the village, the sound of the wind seemed to change. From a breathy, whippy whistle, it had begun to howl, like a pack of wolves, sea-wolves. I shivered, quickening my pace through the empty streets.

9

The Winds of Memory

For much of the night, the strong winds rattled around my hotel and once or twice I heard something clatter, perhaps a chair from the terrace. When I'd booked my room, the cheery hotelier had adopted a surprising approach. 'I only have one room free, but it is not a great room, no view, and you need to be very slim to use the ensuite facilities.' I didn't care about the view since I planned to be outside most of the time, but squeezing between the shower cubicle and the sink to reach the toilet was indeed a tricky manoeuvre. Despite that, and a lack of sleep, I scrubbed away my bleariness and stepped out into the half-dark at 6.30 a.m.

I love the quiet of the early morning, and all of my adult life I have never been able to sleep in, have a long lie. That habit began on a January morning in 1963, the same year when I told myself to catch a grip. Perhaps there was a connection. Through some mysterious process – there was no interview or even any contact before I started work – I landed the plum job of delivering the Store milk, 'the Store' being the local Co-op. My mum worked in the ledger department, and perhaps her hidden hand had been at work. It paid 7/6d a week, a fortune.

One frosty Monday morning, the alarm crashed me awake at 5.30 a.m. and, clutching a slice of toast, I found myself

209

walking through the waking town to the Store milk depot. Turning up was vital, and there was no hesitation in getting up, no matter how cold or how tired I might be, because if the milkman, Tommy Pontin, had to do the round on his own, the milk would be late: breakfasts would not be had, apologies would be endless, and he would be furious.

I liked walking through the darkened streets. I remember once seeing a woman at a high window brushing her long hair as she looked out over the winter morning and, seeing me walking below, she smiled and waved.

Each morning, as Tommy clicked on the electric float and eased out of the depot, I met Jim McCombie, a schoolmate and neighbour. He delivered fresh morning rolls for the bakers, who had been at work since at least 4 a.m. I swapped a pint of gold top for four rolls, and as we did the first few houses Tommy and I munched them. I have been a hopeless addict to Scottish morning rolls ever since.

The float made little noise as it whirred slowly along the darkened streets. With only the clink of bottles and Tommy's incessant whistling, we seemed to fit into the early morning. The sleepy town woke up as we dinked a pint of milk down on its doorsteps; when the float glided past, a patchwork of lights clicked on in kitchen windows and the day began to stir.

Perhaps milk was delivered from St Coombs Farm to the villagers on Lindisfarne in 1963, but in the half-dark of the morning the streets were deserted. The Marygate runs roughly east to west, and when I passed I could see dawn creeping over the North Sea horizon, the sun not yet up but its rays lighting the underside of high clouds, as strong winds chased them across the sky. A door closed somewhere, and out of the corner of my eye I saw someone in a yellow anorak turn quickly into the close leading to St Mary's

Church. A woman, by her walk. I followed the track around the graveyard, wanting to see if I could get across to Hobthrush, the little tidal rock offshore known as St Cuthbert's Isle. The remains of a small chapel had been unearthed and what might have been St Cuthbert's oratory, literally a place to pray, and in the seventh century usually little more than a stone enclosure open to the sky and high enough to screen out the world around. A tall wooden cross had been planted on the islet, but there was still too much water in the narrow channel for me to reach it. And so I decided to climb the Heugh once more to watch the sun come up.

On the summit the wind was vicious, carrying grains of sand that stung my face and neck even when I sat down on the sheltered bench in the lee of the Lantern Chapel. I pulled up my hood to keep out the worst of it. In precisely the same place, for the same reason, Cuthbert and his brother monks will also have pulled up their hoods. In strong winds, the candles in their draughty stone chapel will have guttered and flickered as they knelt to their morning devotions. These moments that reach across millennia touched me, reinforcing the notion that Lindisfarne was like an open history book.

Down at the harbour the morning was stirring. I saw the headlights of a pick-up, and two men unlock one of the upturned boats used as a store. Far out to sea a curved sliver of sun showed over the grey horizon and in moments its first rays caught the topmost towers of the priory ruins. Very quickly, a warming yellow light flooded the island.

Piercing the shrieking wind, another sound came ringing down the ages. High in the tower of St Mary's a bell was tolling. It was 7.30 a.m. I remember being surprised when I saw a notice advertising a daily service of communion at 8 a.m., every weekday. For fourteen centuries, the worship of God had not ceased in the place where Aidan founded his monastery. On the far horizon, a long bank of cloud hid the morning sun, and when I stood up to move off the Heugh the wind pushed hard at my back, forcing me to steady myself. In case they blew off, I took off my glasses and wiped bleary sand-stung eyes. The inscription spoke truth: Wild Lindisfarne.

To my surprise, I met small groups of people, twos and threes, bending against the wind, on their way through the village to communion at St Mary's. In the close, I asked two women if I could watch the service, sitting at the back of the church, and they nodded, smiling, without breaking step. And so, unexpectedly summoned by bells, I sat down in a pew and the service began almost immediately. While the gale raged outside, a congregation of fourteen souls, none of whom had taken off their coats or anoraks, sat in the eastern end of the church, near the altar, and with her back to me a lady in a thick purple coat gave readings and led the opening prayers. Up on the summit of the Heugh, the foul weather had driven out all my efforts to settle, and instead I found myself in the peace and comforting rhythms of a Christian service, much moved by the quiet piety and constancy of a handful of hardy people who sat murmuring the responses to prayers.

Once the reader had closed her Bible and sat down, a priest stood up. I hadn't noticed her when I came in but she was a striking figure for whom the word 'venerable' might have been coined. Small, stooped and with a shock of thick, white hair, the priest wore a green cope over a white surplice that had the Chi Rho symbol embroidered on the back. The first two letters of Christ's name in the Greek alphabet, they had been stitched in the style of the *Lindisfarne Gospels*. Not having an order of service, and knowing little of the rituals of the Church of England, I did not have much sense of the detail of what was happening, but the tone of it was unmistakable. Even though many of the congregation were probably islanders (the tide was still shut) and came to morning communion often, it did not look like a daily routine; instead, there was real warmth and engagement. After a recital of what might have been the Nicene Creed, everyone shook hands with everyone else.

When the priest moved towards the altar, she turned to face the congregation and, in a gesture that provoked an ancestral shiver (to which I had no right, as a failed Scottish Presbyterian), crossed herself. With the reader, who appeared to have a role as an assistant, the priest prepared the wine and the white wafers for the ceremony of communion. Once all had been blessed and the rumble of the Lord's Prayer subsided, the congregation knelt at the altar rail to receive the body and blood of Christ. In the past, I used to wonder at the literal nature of this ritual, with its prayer about eating 'the flesh of your dear Son Jesus Christ', but on that stormy morning it moved me very much. Faith is far from simple, I suspect, but in this warmly lit little church it seemed like both a bastion and a refuge from the tumult of the world.

Once all had resumed their seats, the priest appeared to eat the wafers and drink the wine that was left. Perhaps having been sanctified, they could not simply have been discarded. And then she blessed the congregation and the service was over. Sitting on a pew at the back of the church, I waited for everyone to leave. For some reason, I wanted to be alone as I lit more candles for the little ones, for Hannah, Grace and Hugo.

When I left the shelter of the porch, the wind had dropped and I walked down to the shore. The tide had retreated far enough to allow safe passage across the sand and rocks to Hobthrush, St Cuthbert's Isle. For such a breezy day, I was surprised to see so many cars streaming across the causeway, but reckoned few would want to spend time on the islet, where there was little to see.

Certainly by 678, and likely long before then, Eata had sent for Cuthbert. The old abbot wanted him to give up his hermetic life, wandering the secret tracts of solitude on the moors and mosses of the Kyloe Hills and beyond, and to re-enter monastic life as Prior of Lindisfarne. Cuthbert's unquestioned piety will have given him authority, something that may have been much needed after the convulsions of the Synod of Whitby in 664. Unwilling to accept that all should conform to the liturgical and political dominance of the church in Rome and the papacy, Bishop Colman and a number of monks withdrew from Lindisfarne to return to Iona, the west of Scotland and Ireland. Despite their ejection from Ripon a decade before for refusing to accept Rome and Wilfred's control, Eata and Cuthbert were left with little choice. The island community had to conform after the king of Northumbria had ruled in favour of the papacy at Whitby.

Both a bishop and an abbot, Eata had a see as well as a community to administer, and during his inevitable absences as he travelled around Lindisfarne's vast patrimony in the north of England and southern Scotland Cuthbert was left to guide his brother monks and sit in judgement when disputes arose. The early sources are surprisingly frank about his difficulties, and here is a passage from Bede's prose *Life*:

Now there were certain brethren in the monastery who preferred to conform to their older usage rather than to the monastic rule. Nevertheless he overcame these by his modest virtue and patience, and by daily effort he gradually converted them to a better state of mind. In fact very often during debates in the chapter of the brethren concerning the rule, when he was assailed by the bitter insults of his opponents, he would rise up suddenly and, with calm mind and countenance, go out, thus dissolving the chapter, but none the less on the following day, as if he had suffered no repulse the day before, he would give the same instruction as before to the same audience until, as we have said, he gradually converted them to things he desired.

Except he may not have desired them, at least not wholeheartedly. Cuthbert had been ordained and taught by Boisil in the Irish tradition at Old Melrose, had worn the Druidic tonsure, and his asceticism and hankering after the hermetic life derived from Celtic rather than Roman impulses. Eventually, and perhaps not surprisingly, these disputes seem to have worn down the prior's patience. Bede noted, 'At the same time he kept a cheerful countenance though sorrows overtook him, so that it was made clear to all that, by the inward consolation of the Holy Spirit, he was enabled to despise outward vexations.' But not endlessly, and it eventually became too much for Cuthbert.

Bede again:

Now after he had completed many years in that same monastery, he joyfully entered into the remote solitudes which he had long desired, sought, and prayed for, with the goodwill of that same abbot and also the brethren . . . Now indeed at the first beginning of his solitary life, he retired to a certain place in the outer precincts of the monastery which seemed to be more secluded.

It was not the first time Cuthbert had sought the solitary life, something Bede probably knew but chose to ignore, and this time he was careful to point out that permission had been granted. This certain place was the islet of Hobthrush. On that windy morning, I picked a cautious way between slippery, seaweed-covered boulders to reach it at low tide. Little more than a small outcrop of black dolerite rock, Bede wrote that it was more secluded, *'secretior'*. Remembering the different meaning of the noun 'seclusion', a simpler translation of the original Latin might be 'more solitary', more deserted, a diseart. The early morning high tide had heaped seaweed on the smooth rocky shelves that led like giant steps to the grassy area where the wooden cross had been planted. The flattish area was not much larger than a tennis court, and to its south-west a dark hump of rock rose like a bulwark against the waves. On it a plaque quoted Psalm 93:

> Mightier than the thunders of many waters,
> Mightier than the waves of the sea,
> The Lord on high is mighty!

By the time I had scrambled up to the cross, I saw that it had been raised at the eastern, altar end of a small rectangle of masonry, the foundations of a medieval chapel that had been dedicated to St Cuthbert long after his death. In the 1880s, Major-General Sir William Crossman, an amateur archaeologist, had unearthed more medieval ruins and near them what he believed to be 'the site of the cell to which St Cuthbert was wont to retire'. This may or may not have been an oratory, and Crossman recorded that it lay in the south-east corner of the islet, but unfortunately I could see nothing of it. Instead I sat on the low wall of the medieval chapel, opposite the tall wooden cross, looking back at Lindisfarne, the Heugh, St Mary's and the ruins of the priory.

Even though the wind had dropped dramatically, there were few who ventured down to the beach and so far none had attempted to cross to the islet. Perhaps I would be left alone with my thoughts, to find some peace in precisely the place Cuthbert had when he fled the rancorous disputes of the monastic chapter. I am not certain what I expected from walking where he walked, perhaps only some time by myself to think about the past and what might be a worthwhile way of looking at the future. I smiled to think that it was not the secret of life I sought, but the secret of death.

Unfortunately, I noticed a couple moving through the sands and rocks towards the islet. They were walking quickly and I could make out a tall man wearing shorts and a woman wrapped in a bright sky-blue parka and matching trousers. Speaking a language I didn't recognise, they quickly scouted the location, taking only moments to look around, and before they left the man came up to the wooden cross to give it a good shake. Maybe he was anxious about how secure it was in this exposed place. Then he beckoned to his partner to take a photograph of him in front of it. And to my open-mouthed horror, he stretched out his arms in an imitation of the crucified Christ. I was appalled, but not being a Christian I could scarcely accuse this lout of blasphemy, though he was certainly being grossly disrespectful. I am afraid I didn't conceal my anger, but before 'Hey! You!' could develop into anything else the couple trotted off back to the beach without a backward glance.

Much annoyed at becoming annoyed, I sat back down on the wall head, but this time looked out to sea and across the southern sands to the Northumberland coast. Its seamarks are dominated by the looming mass of Bamburgh Castle. Heavily restored by the wealthy industrialist William Armstrong after 1894, it is a magnificent fortress perched on a large, steep-sided dolerite outcrop. It was rarely captured by force and for centuries was the capital place of the kingdom of Northumbria, a fact announced by surprising roadside signs outside the tiny village.

In Cuthbert's time, a wooden stockade crowned the great rock, and inside it timber halls, long lost under the medieval cobbles and Armstong's rebuilding, housed the royal retinue. As the winds whipped the waves of the North Sea and howled around their walls, fires blazed, warriors feasted and bards sang of the power of the kings and their queens. Bebbanburh is the earliest recorded Anglian name for Bamburgh and it remembers Aethelfrith's queen, Bebba. The summit of the rock was large enough to accommodate hundreds of people: servants and grooms, and soldiers to defend it. When Aidan chose to found his monastery on Lindisfarne, his decision was much influenced by secular as well as spiritual considerations. The saint's mission to convert and re-convert the Anglian population of Northumbria needed royal support that was close at hand.

Aethelfrith's spectacular gains in the north after his victory at Addinston in Upper Lauderdale in 603 were matched by expansion to the south. When Ida first took over the British fortress of Din Guauroy, the older name for Bamburgh, Aelle was carving out the kingdom of Deira. Like Bernicia, the name is from Old Welsh and derives from *derw*, 'the oak tree'. At first Deira extended from the mouth of the River Tees down the Yorkshire coast as far as the Humber. Some time after 603 Aethelfrith annexed Aelle's young kingdom to become the first ruler of Bernicia and Deira, what was called the land of the Northanhymbra, the people north of the Humber. It grew into the most powerful of all the emerging Anglo-Saxon kingdoms and its rulers were the first to call themselves Bretwaldas, 'Britain-rulers'.

In 616, after the death of Aethelfrith in a battle fought near Chester against the Welsh kings, Aelle's son, Edwin, engineered a reversal, the takeover of Bernicia by Deira. He had also conquered the British kingdom of Ebrauc based at Eboracum, York, and begun to adopt Roman forms and customs. The great basilica of the imperial city still stood and there is evidence that it was used by Deiran kings. When he made his royal progress around his expanded kingdom, Edwin basked in the borrowed authority of Rome when he had the imperial symbol of the tufa carried before his retinue. On Edwin's death in 633, Oswald came back from exile in the Celtic west of Scotland and reasserted Bernician claims to the twin kingdoms, and it was he who invited Aidan to come from Iona to Lindisfarne. Until the early decades of the eighth century, direct descendants of Aethelfrith and Aelle were kings in Northumbria as it grew in power and its reach extended over much of northern and western Britain. When Cuthbert retreated to Hobthrush and looked out over the sands to Bamburgh, he knew that his monastery had been deliberately placed at the centre of political power.

Thinking about the balance between sacred and secular, I found myself guilty of assuming a modern dichotomy. We have long looked at Church and State as linked but separate, but of course Cuthbert and the kings in their halls at Bebbanburh made no such distinction. God was not compartmentalised and the role of priests limited to Sunday services, funerals, baptisms and weddings. Instead the powers of Church and State worked together, as Northumbria expanded in all directions. Conquest and subjugation were God's will, the work of His Church, and the bishops of Lindisfarne understood that their role was deeply political. Most prominent churchmen were, in any case, aristocrats. All of this bore in on Cuthbert as he argued with his brother monks, trying to persuade them to abandon Irish Celtic practices and conform to the teachings of Rome, and especially to the uniform dating of the great festival at Easter, something disputed only by the king's enemies. All had to be united in pursuit of Northumbrian glory. If kings succeeded, so did God, and great crosses were raised to His glory at Ruthwell, Bewcastle and in other conquered territories. Even though the native British believed themselves to be devout Christians – the Baptised – there is a sense of conversion to the true faith in the process of Northumbrian expansion.

Well aware of all of this, Cuthbert was beginning to withdraw when he came to Hobthrush, built his oratory and began to pray, shivering in the sand-stinging winds and the cold rain. But he did not stray far from the centre of power, not then and not later in his hermetic life. In trying to look at the past like a competent historian, trying to put myself physically and emotionally in the mindset of people in the past, it occurred to me that Cuthbert might have been ambivalent. He dearly wished to flee the world of dispute and raw politics, and focus all of his will and self-denial on making certain of his place with God in heaven, but not entirely. Hobthrush was, after all, like Lindisfarne, a tidal island. Perhaps somewhat conflicted, he recognised that in himself, and set up his hermitages close to the centre of power and not in the wastes beyond the Kyloe Hills or even further afield. Turning this over as I sat on the wall head was comforting. Cuthbert's devotion and bodily privation were real enough and a world I could never enter, but his doubts and unwillingness to leave the world of politics brought him closer, made him appear a little less saintly, less remote. God was nearer on Hobthrush, but so were the halls of Bamburgh.

Needing to move and warm up, I walked around the

perimeter of the little islet, and only a few yards out to sea two seals were playing. Their dog-like heads bobbed above the surface and one seemed to flip on its back before diving under. Before I returned to Lindisfarne, I did something that surprised me. Without thinking why, I stepped back into the rectangle of the small chapel and, looking at the wooden cross, I felt myself begin to think about some long-held responsibilities.

None of my three children asked to be born; Lindsay and I wanted to have a family and we decided to bring them into the world. That bound us with welcome and wonderful ties of a love I had not experienced before, and all my life I have known that I would die to save any of them. But now Adam, Helen and Beth are all in their thirties, adults who have made their own way, have taken their own decisions, and while we will offer any and all sorts of help they are no longer our responsibility. Soon Grace will have cousins, we hope, and their lives all wait to be lived. Now, I realised, I had come to this beautiful place to begin to learn how to leave them, to learn how to live what remains of the rest of my life and how to die when the time comes. On this little scrap of rock, Cuthbert had been doing something similar: beginning to leave the world and learn how to die. And so in front of the cross, very awkwardly, I spoke to him – out loud, my words blown out to sea by the wind. Like him, I would begin to spend more time alone, not working or writing but thinking, talking to myself, confronting honestly the darknesses of past wrongs, failures and regrets and accommodating them. If I could do that, then what is left might just be happier, saying goodbye to Lindsay and my children easier to bear. Across fourteen centuries, it seemed as though a long-dead saint was becoming my soul-friend.

Once across the seaweed-covered rocks and the sands of the channel between Hobthrush and Lindisfarne, I decided I needed to be in the company of people, just be with them and not talk. Another visit to Pilgrims Coffee House and its predatory sparrows seemed like a good idea. With a fruit scone the size of a side-plate, I went out into the garden but, the tide not being shut, all of the tables were taken. Then a couple got up and I joined an Australian lady at a small table. Despite my resolution to be quiet, I asked if she had been to the island before. When she surprised me by saying she came every eighteen months or so, I asked what brought her halfway across the world to this place. She smiled: 'This is where I keep my soul.' And when I asked what she did on Lindisfarne: 'Walk. I just walk every day around the island.'

By the time I had fought off the sparrows and eaten my scone, the tide was shutting and I decided to emulate this lady's habits. Fortified and under a bright sun, I made my way out of the village and around the bay towards the castle. Below it stand a series of large limekilns, places where limestone was burned at very high temperatures to produce fertiliser and a key ingredient for builders' mortar. Leading north along the eastern shore of the island is an old waggonway that supplied the kilns. Flat and high above the stony beach, it made for easy walking. Not far from the castle, the tides had pushed up a long ridge of stones and on it were a series of small cairns, like the crenellations of a castle rampart. Closer-up, they were clearly carefully made and not like cairns at all. Some had much larger stones balanced on smaller ones, a little like the pillar at St Cuthbert's Cave. Others had small stone circles around them, and still more were artistically shaped, almost sculptural, decorative. Several had collapsed. I suspect they have a simple function: people who visit want to leave a mark on this sacred landscape, like the plaques on the benches. I liked them.

When the waggonway climbed a little, I could see that the wind was driving spectacular breakers as the tide flooded in. Sandbanks were pushing some of them in different directions, causing them to collide before they crashed onshore. This made surging, sideways runs of waves that seemed to suddenly die away. It was hypnotic.

Many years ago, I saw the same phenomenon in Uig Bay on the Atlantic shore of the Isle of Lewis in the Outer Hebrides. I had gone to see the ruined township of Carnais, one of the many that had been abandoned during the dark times of the Clearances, the forced evictions of crofters that took place in the nineteenth century. What brought me to this heart-breakingly beautiful place was a song. 'An Ataireachd Ard' is a moving, soaring and deeply affecting lament. It speaks of the suffering and loss of whole communities and their way of life, but also the wash of history across the landscape. It could as easily have been written on Lindisfarne as on the Isle of Lewis.

On a summer afternoon in the 1890s, Donald MacIver hitched up his pony and trap for a journey back into the past. A teacher at the school at Breascleit, he had received a letter from Canada, from his uncle, Domhnall Ban Crosd. Forty and more years before, he and his family had been cleared off the land at Carnais and herded on to the emigrant ships bound for North America. But before he died, Domhnall Ban wanted to see his home-place once more. Many who were forced into exile suffered something more than homesickness; the Gaelic word is *ionndrain* and it means something like 'a missing piece', perhaps a part of the soul that had been left behind on the windy shores of the Hebrides. As his nephew clicked his pony up the track at Miabhig, the vast panorama of the mighty Atlantic opened before them and the pale gold of the sands of Uig Bay glowed in the summer sunshine.

Domhnall Ban's nickname of 'Crosd' is a Gaelic version of the English word 'cross' or 'grumpy', and as he saw again a place he had beheld only in his dreams the old man's stern, stone face was set, betraying no emotion. But when at last they rounded the bay and came to Carnais, the place where he had been born and raised in the body warmth of an interdependent crofting community, there was nothing to see but ruins, only a few courses of tumbled masonry where houses had once been. There was little left of the busy, working landscape where families had toiled, lived and died. At last Domhnall Ban's face crumpled and he wept for the loss of his home, the image he had held in his thoughts across all of the years of exile in Canada, and he was desolate at the waste. *'Chan eil nith an seo mar a bha e, ach an ataireachd na mara,'* he said to his nephew: 'There is nothing here now as it was, except for the surge of the sea.'

Much moved by the memory of the old man's tears, Donald MacIver wrote his great lyric about loss, change and the tides of history:

> *An ataireachd bhuan*
> *Cluinn fuaim na h-ataireachd ard*
> *Tha torunn a'chuain*
> *Mar chualas leamsa nam phaisd*
> *Gun mhuthadh, gun truas*
> *A' sluaisreadh gainneimh na tragh'd*
> *An ataireachd bhuan*
> *Cluinn fuaim na h-ataireachd ard.*

The ceaseless surge
Listen to the high surge of the sea
The thunder of the ocean
As I heard it when I was a child
Without change, without pity
Breaking on the sands of the beach
The ceaseless surge
Listen to the surge of the sea.

As I walked along the waggonway watching the crash of
the waves, the rhythm of Donald MacIver's words in my
head, I realised that Cuthbert saw exactly what I saw and
heard the eternal thunder of the ocean exactly as I heard it.
Much moved by that continuity, at last I began to confront
my own history. Though I hoped that the sadness that
engulfed Domhnall Ban Crosd towards the end of his life
was not waiting for me.

When I reached Emmanuel Head, a vast white obelisk on
the north-east corner of the island that acts like a daytime
lighthouse, keeping shipping from setting a course too close
to the reefs and sandbars that lie offshore, I found a bench
and sat down to stare at the waves.

Resentments fade and anger cools as the years race past,
and when I look back at my time in television, the failures,
problems and betrayals seemed to have diminished with
distance. In that liminal time in the early morning, in the
grey wastes between waking and sleeping, thoughts of what
happened twenty years ago do not often drift into my head.
More recent history does, however, take up too much
emotional energy and since I probably don't have twenty
years left to allow the pain of it to pale, it occurred to me
on that blowy afternoon that I should try to come to a
settlement sooner rather than later.

Sitting on a bench, looking out to sea on Cuthbert's island, I felt that by allowing those who had hurt and attacked me to intrude, I was somehow polluting this place of peace. But in truth I needed to deal with this. In St Mary's Church, only a few hours before I sat down, I had heard the murmur of the Lord's Prayer, the words so familiar that I didn't think about them. One line came floating in the clear air at Emmanuel Head and that was 'forgive those who sin against us'. I knew that was what I had to do – forgive these people. I knew I had to do that if these painful episodes were to be banished to the margins, but I had no idea how.

The shadow of the white obelisk at Emmanuel Head was beginning to reach across the sea as the westering sun slid behind it and I decided to walk back through the evening to the village. I noticed that, like most benches, there was a metal plaque on the back of this one. It commemorated Gisela Elfriede Hall (née Rohde) and under her name was an inscription in German: *'Hier war ich immer glucklich und zufrieden'* ('Here I was always happy and content'). Gisela's surname suggested that she had married a British national and perhaps she came often to the island. In any case, the inscription encouraged me to think that Lindisfarne might indeed calm and console me, if I let it. It also reminded me not to waste time. Gisela died in 2011, only a few days short of her sixty-eighth birthday. She was younger than I am now.

On the waggonway, I had been startled by some very striking and large wickerwork sculptures of two adult ducks and their ducklings set in the marram grass by the beach. For a moment I was not sure if they were animate or not. As I turned inland from Emmanuel Head, I came across more of these playful works of art: a gigantic fly anchored to a sign, a huge owl perched on a post, and a tall broadleaved plant that looked tropical attached to a fence.

I was making my way to another track that led down the centre of the island past St Coombs Farm and on into the village. These are called lonnen, a variation on the Scots word 'loaning', a green track, usually between fields. It was the only place with significant tree cover and my walk was well shaded.

I had seen a black horse grazing amongst a flock of very well-covered sheep and, as I passed its field gate, a woman waited. She had caught up the horse in a halter and did not want me to walk behind it. She was worried it might spook, or plant its feet and refuse to move. I walked ahead and we talked briefly before she turned her horse out in a paddock with a field shelter. Only about one hundred and forty people now live on the island, she said, since so many houses in the village had become holiday lets and there were only two children attending the school, one of them part time. When I told her I had been walking, she smiled and said that was how she found peace, just by walking and sometimes riding along the island paths.

A dusty light was fading by the time I reached my hotel. After shuffling sideways into the shower and enjoying more Holy Island gin, I walked to the Ship for another Reading Tea. The restaurant being fully booked, I ate at a table in the bar.

After a day of sun, clear skies made it a moonless night, and even though I was tired I took myself up to the Heugh to look out over the sea. The high winds of the morning had softened to a breeze blowing out of the south and the starlight shimmered over the water. There was enough light to see the castle in silhouette; the lighthouse on the Farne Islands seemed far out to sea and very brightly lit, and Bamburgh Castle looked warm and welcoming.

Well wrapped and with many calories from the Ship Inn

supplying central heating, I found myself going over the events of the day. There seemed to have been many – a long time had passed since I sat on the Heugh at dawn and went to communion at St Mary's. I had read and heard that people loved this island with a quiet passion and I was beginning to understand rather than intuit why and how. They walked, beyond the village and the press of visitors – that was where the essence of this place was to be found, where peace could descend. Cuthbert walked and sang psalms or recited prayers as he passed the crash of the waves or heard the cry of seabirds. Even though the priory ruins, St Mary's, the castle and the other attractions in the village supply the images that make up Lindisfarne's famous iconography, it was beyond these that some sense of its secrets could be discovered. And that was what I would do for most of the rest of my stay – walk.

In the weeks before I crossed the causeway, I had read about and become interested in religious visions, not because I imagined for a moment that I might have one on Lindisfarne, but because they seemed to occur in Cuthbert's time. I wondered if these leathery old saints, fasting and up to their necks in freezing water, had had hallucinations and reckoned them to be visions, revelations of the sort Drythelm recounted. These were understood as events, blinding flashes or dramatic turning points that could change lives and the perceptions of those who listened. One of the most famous visions of the early modern period came to an entirely rational, brilliant mathematician. Blaise Pascal invented the mechanical calculator and made important discoveries in geometry and arithmetic, yet in 1654 he had a profound religious experience that changed his life. It was so central that he wrote this on a small piece of paper:

FIRE. God of Abraham, God of Isaac, God of Jacob, not of the philosophers and scholars. Certainty. Certainty. Feeling. Joy. Peace.

So that it would always be with him, Pascal had the paper sewn into the lining of his coat. And if he changed his coat, this brief record of what he experienced was transferred. After he had seen God and felt joy and peace, the mathematician wrote *Pensées*, his *Thoughts*, and before I came to the island I read some of this beautifully written text. The sincerity of his belief and the certainty of what he had seen were unmistakable. And it was a moment after which everything was different, something I knew I was very unlikely to experience. Instead, I suspected my own search for peace would be a process, one that might take time. But having no belief in God or any other agency, it is what I think that matters, and I was now sure that I had to do it and I reckoned I had made a start.

The wind had dropped, but when I walked through the darkened village to my hotel I was certain I could hear it howl once more. This strange, chilling sound seemed to come from the west, the sands beyond the island.

10

Duneland

It was an open sky as I scurried down the Marygate to the harbour. Dawn would come up clean and clear, and I wanted to see the sun rise over the North Sea horizon. No clouds would occlude it and, seen from the castle, the highest point of the island, I hoped the coming of the morning would look spectacular. But I was late, not out of the hotel until 6.45 a.m. and sunrise was due at 7.01 a.m. Like an early morning jogger, I trotted through the half-light as quickly as my creaking knees would allow and climbed the cobbled path to the castle entrance and the terrace beside it. The moment I stopped, it started.

Every time I have seen a sea sunrise, the drama of it has been breathtaking. At first the glinting edge of the fiery disc is just discernible and it is possible to look directly at it without being dazzled. And then it happens with a speed that always surprises me. When the top third of the sun showed, it began to light the battlements above me and then it flowed down to warm me on the high terrace. I looked behind and the long shadow of the castle reached as far as the towers of the priory and the roofs of the village below it, perhaps three-quarters of a mile away. All of the colours of the land began to glow: the green of the fields, the russet of the autumn trees that line the lonnen, the brilliant white dots of grazing sheep.

So that I could watch the sea change colour from a cold grey to blue-green, I left the terrace to climb up to Little Beblowe, an outcrop directly to the east of the castle. A rabbit scuttered in front of me, startled to see movement so early, and when I had scrambled to the summit, the sea and the land seemed to come alive. On the far horizon, the hull of a tanker glinted; to the south, the ramparts of Bamburgh were warmed; on the mainland, the low sun lit the silent stream of traffic on the A1. Darkness had fled, the light had come and the day could begin. A random thought swam into my head. How sad it was that Lucifer, the Bringer of Light, Son of Morning, had fallen and become the symbol of evil. Mornings like this lift the heart high.

On my way back to the village to light candles at St Mary's, the scent of a summer just passed filled the air. On a patch of lush grass by the harbour road, an old man was quartering the ground on a ride-on mower, making contrasting light and dark green stripes with the roller behind the blades. Without stopping, he lifted a hand in greeting and I walked up the Marygate, its houses basking in the butter-coloured sunlight. The first frosts would come soon enough and the trees were already in the yellow leaf.

The line of the Marygate is thought to run along the northern edge of the monastic precinct, the ditch and bank that marked off the most sacred ground. I had become curious about routes around the island, the clear link between Cuthbert's habit of walking while he sang and prayed and those pilgrims who came now. The village and the medieval priory have obliterated the precinct, but there are fascinating survivals that offer a partial sense of what it was like and how it was seen. No crosses like those at Ruthwell and Bewcastle survive from Lindisfarne; there is only the massive socket between St Mary's and the priory

and what is known as the Petting Stone. This sits to the left
of the path leading to St Mary's and was the base of a
monumental cross. When a rare wedding takes place in the
church, the bride takes part in a fertility rite as she jumps
over the stone, helped by the two oldest fishermen on the
island. Another is tucked away in a corner of the medieval
ruins. It has carvings that date to the eighth or ninth centuries and was recovered only when the central tower fell down
in the 1820s. The medieval builders had split the old cross
socket in two and used it as foundations for the tower. It
seemed to me a pity that more was not made of the two
sockets: they are the sole direct survivals *in situ* from the
original monastery.

But in the priory museum there is an excellent display of
other survivals, what the curators call 'name stones'. They
looked to me to like small tombstones. Fourteen have been
found around the priory and the style of carving suggests
that they were made between about 650 and 750. Cuthbert
would have seen them and perhaps known some of those
who were commemorated. The stones were brightly
coloured with reds, greens and black painted over a white
background and the lettering of the names and the style of
the crosses carved on all of them definitely relate to the
Lindisfarne Gospels. What made these stones come alive for
me were the names inscribed on them. Beannah, Osgyth,
Ethelhard, Aedberecht, Audlac and the others may have been
monks or individuals privileged or sufficiently wealthy to be
granted burial inside the monastic precinct. It is a very early
example of a cultural habit that persisted into the modern
period, the wish to be buried in holy ground, something
seen at Dryburgh Abbey with the tombs of Sir Walter Scott
and Field Marshall Earl Haig.

It may be that the earliest cross was erected after the death of Cuthbert in 687, and it is believed that when the monks left the island in 875, at the approach of the Great Heathen Army, they took it with them. I wondered if there were other crosses or memorials of other kinds beyond the precinct, places that could be visited much in the way that the stations of the cross are in Jerusalem. Forty years ago, I did some consultancy work in that city and took time out to follow this remarkable procession. One of the many elements that made it powerful was that it led to such an immense climax, one that clearly affected people profoundly. Perhaps there was a similar route on Lindisfarne that took people around the island and eventually brought them to Cuthbert's cross and then his shrine, the place of greatest sanctity. And where did Cuthbert himself walk? As ever on this journey I wanted to follow as closely in his footsteps as I could.

Even before he retreated across the rocks and the seaweed-strewn sands to build his oratory on Hobthrush, I suspected Cuthbert had already begun to go there to seek peace and relief from the bickering with those brethren who would not conform to Roman practices. And so I decided I would start my search for the saint's secret places there. Once I had clambered up the slippery-stepped approach, I realised I was not alone on the islet. Sitting on the low wall head of the ruined chapel, almost exactly where I had been the day before, was a young and very beautiful dark-haired woman. Her hands stuffed into the pockets of her anorak, she was looking out to sea. When I greeted her, she looked up briefly and smiled, saying nothing. Having always found it difficult to resist conversation, especially with a woman so good-looking, and thinking this situation oddly awkward, I asked her about the howling noise I had heard each night. Without standing up, she turned and pointed to the sandbars south of the causeway. I could see that one was covered with many black shapes, some of which seemed to be moving. 'It's the seals. At night they beach on the sandbars and call out.' I thought of the selkies, the captive seal-folk in the legends of the Northern Isles who long for the sea.

The woman also told me that she came alone to Lindisfarne every two or three years because 'I feel at home here'. And then she stood up, smiled again and began to make her way off the islet. Later that day I saw her in the queue at Pilgrims Coffee House and we simply smiled at each other without exchanging a word. For her and several others I saw walking alone, who appeared to be by themselves, their visits to the island seemed not to be sociable in any conventional sense. They had come to talk, but only to themselves, and perhaps with their god. Leaving aside the latter, I was the same, realising that I too had joined the community of the uncommunicative.

The Pathfinder map of Lindisfarne shows two wells. Marked by some old and rusty winding gear, the Popple Well lies near the foot of the Marygate and was probably inside the monastic precinct, while the Bridge Well is on the lonnen that ran from the eastern coast of the island to St Coombs Farm, what is known as the Crooked Lonnen. I decided I would walk north out of the village towards the Links, a large area of dunes, and the place where my friends and I had fled to in 1965, and I would come back by way of the Bridge Well.

Almost from the moment I left the tarmac road at Chare Ends, the landscape looked familiar. Over the span of fifty years, trees grow up, buildings are built and perspectives shift, but nothing much had changed since 1965 as I walked up the western lonnen past the fields of the farm. I remembered the half-buried ruins of limekilns, although at the time I had no idea what they were, briefly considering them as good places of concealment. There seemed to be ridges of sand dunes, and between them wide flat areas of grassland with obvious paths leading through them to gaps in the next ridge. The term 'links' is now usually applied to golf courses, but it derives from an Old English word, *hlinc*, cognate to flank. It means 'marginal land'. Cuthbert probably knew and used the word. The morning was very bright and, in the shelter of the high dunes, walking was warm work. I had read of one visitor spending a day 'holed up in the dunes with a flask of tea and a book'. It sounded good, but I needed to march on and discover more about the shape and nature of this part of the island, so different from the detail and domesticity of the farm and the village.

Between two ridges of dunes I came across the ruin of a single-storey building with a gable end still standing at one end of long, low rubble-built walls. I later discovered it was marked on nineteenth-century Ordnance Survey maps as the Shiel. Probably a shorter version of shieling, a place where shepherds lived while they summered out with their flocks, it looked old to me. Its shape suggested a medieval longhouse, and perhaps it had been altered and adapted over centuries of use. The walls still standing were wonderfully well-built, each stone keyed into the next to make a strong bond. Mixed with a good deal of yellow sandstone was some grey whin, the same stone the Heugh is made of and the castle stands on, and one or two stones carried some lovely pink veining.

In the summer of 2017 a tall wood of mature sitka spruce and some majestic Scots pines that stood north of our farm-house and sheltered the home paddocks was cut down. It belongs to my neighbour, and while I was very sad to see the Scots pines go, the scruffy sitka were being blown down by winter winds. The machinery that arrived to cut the trees was awesome in its power, the harvester taking only seconds to saw through tree trunks and sned off all of its branches before dumping it on a stack. With its swinging saw-head, it looked like a dinosaur browsing the primeval forest. The churn of the harvester and forwarder on the edges of the wood unearthed something interesting. There used to be a drystane dyke between it and our track, and one of the foundation stones, a bedder, had been pulled up to the surface. It was beautiful: dark sky grey, with veins of pink quartz running through it. It is characteristic of our valley, and many of the bigger stones in the farmhouse walls have similarly vivid veining, like the longhouse on Lindisfarne. Whin was picked up off the fields and most of the older houses are built from it. Because it is hard and difficult to work, the walls are built drystane style, although old-fashioned lime mortar is used to bind them. This means that, with the exception of gables and corners, only one face has been completely squared and a lot of small packing stones have been used to create level beds for the next course. This is why the walls are very thick. They have a rubble core packed between the inner, undressed ends of the whinstone. It is very beautiful, and this sort of construction takes great skill and a good eye. When I look at the corners of the old farmhouse, they stand straight as a die. And I could see evidence of the same sort of work on the ruins in the dunes.

When I was a student, I worked many of my summers and Easter holidays as a builder's labourer. Most memorable were several weeks working for two old stonemasons (well, they were a lot older than I was at the time) who were building a high retaining wall at the entrance to a new housing development off the Castlegate in Jedburgh. Willie

Hinnegan worked with Johnny Ferguson, and my job was to lift up stones on to the scaffold so that, with their mels and chisels, they could cut them to size. Often they turned a stone over and over, and with only a light tap would lay it open. We were using rubble from a stretch of drystane walling that had been demolished (no doubt too expensive to maintain) and brought to the site. Johnny and Willie would point to one, 'See's that bonnie merkit yin', and as the wall grew higher, so did the scaffold. Johnny Ferguson was also known as Johnny the B because he could be one. And after he had booted three big and heavy stones off the scaffold in ten minutes, in what seemed to me to be a clear attempt to wind me up, I told him that if he did that again he would be joining the rejected stones. Willie calmed us both down and we got on with the job. The wall is still there and has grown more beautiful with age, bonnie merkit.

Near the ruins of the longhouse was a high sand dune and it looked familiar enough to persuade me to climb it. It had panoramic views south to the castle and the village and north and east to the sea, and even though it was flat-topped I could not be sure if it was where we hid ourselves in the summer of 1965. In any case, dunes shape-shift a good deal in the wind. From the high vantage point, I could see most of what I had come to call the duneland. And even though it was a clear, sunny day with little wind, I realised I was alone. No one else walked the path I had come, or any of the others that threaded through the sand and the marram grass. The tide was not shut, cars had streamed across the causeway, but it seemed that no one had ventured beyond the village, the harbour or the castle. When Cuthbert walked the island, the population was tiny, perhaps only a few dozen brothers, their servants and some children being taught at a school established by Aidan. And so when he came to the duneland, longed-for peace could often be found.

At that moment, I realised that this empty place will have meant much more to Cuthbert. Surrounded by silence, solitude and sand, he was in the deserts of St Anthony of Egypt. When the sun beat down, such as on a day like this, Cuthbert could have imagined himself walking in the long shadows of the Desert Fathers. Emulation and example were particularly important means of cementing faith for early Christians, and a man born and raised in the green hills and grassy meadows of the Tweed Valley found himself not only where St Anthony had walked but also where Christ had been tempted three times by Satan. And just as on top of the great rock of Cuddy's Cave, Cuthbert could have climbed the dune I stood on and seen the royal stockade at Bamburgh and all of the kingdoms of the world. Even more than that, he could turn north and east to look out over the empty wastes of the sea and know that his desert, this *diseart*, was doubly protected by God. The sea, the sands and solitude made possible by Creation meant that the duneland and the whole island was a place where God could see those who loved him.

Beyond the longhouse and its sheltering dunes, I walked down to a long sandy beach, the sort of place that would usually have been crowded with sunbathers, sandcastle builders, splashers in the waves and strollers. But there was no one. I could see for more than a mile from west to east and I was the only person on this beautiful, blessed beach. When Cuthbert came here, no doubt in all weathers, the glories of this place may have inspired him to sing from the psalmody. Biblical in origin, many of these were known by early British Christians, and Old English versions of Psalms 22 and 23 exist. The latter is perhaps the best known of all,

'The Lord Is My Shepherd', and even as a non-believer I can still recite the words and sense its power. The Lord was Cuthbert's Shepherd, and on that crystal day I imagined him singing it:

> The Lord's my Shepherd, I'll not want;
> He makes me down to lie
> In pastures green; He leadeth me
> The quiet waters by.
>
> My soul He doth restore again,
> And me to walk doth make
> Within the paths of righteousness,
> E'en for His own name's sake.
>
> Yea, though I walk in death's dark vale,
> Yet will I fear no ill;
> For Thou art with me, and Thy rod
> And staff me comfort still.
>
> My table Thou hast furnished
> In presence of my foes;
> My head Thou dost with oil anoint,
> And my cup overflows.
>
> Goodness and mercy all my life
> Shall surely follow me;
> And in God's house forevermore,
> My dwelling place shall be.

Even though these beautiful words will have been blown out to sea in the swirling wind, God heard them.

At the eastern end of the long beach a rudimentary fence

and some black-and-yellow striped tape suggested I move inland. Beyond the barrier, the going did look rocky, and very narrow, even when the tide was out. With some difficulty, and careful to do nothing that would restart my back problems – what I had come to call Lindisfarne Leg – I climbed slowly up a very sandy dune, cascading small landslides with each step, and after a struggle found myself on a plateau of thick marram grass. Beyond it, I could see the Back Skerrs and the rocky shore of Coves Haven. The sweep of this bay and the high cliffs of Sandham Bay beyond it are of a different scale, much more dramatic than the shoreline around the village and the priory. On the beaches of each one I found some twisted shapes of driftwood, what must have been a welcome gift of firewood for the monks; they were sometimes so contorted that for a moment I thought one was the skeleton of a cow. At intervals in the duneland there were small rubbish dumps, places where people had gathered up plastic bottles and all sorts of other debris washed up by the tides. In the distance I could see Emmanuel Head, the white obelisk made brilliant in the sun. And when I reached the daymark, as sailors call it, there was company of sorts. Waiting for the breakers were two surfers and they seemed very skilled, pulling themselves upright on their boards at just the right moment to ride the waves. Their heads just above the water, three seals were watching them.

The daily isolation on Hobthrush, his walks through the duneland, where he also kept night-long vigils, did not satisfy Cuthbert's longing for the hermetic life. Some time in the early 680s, and perhaps before, he persuaded Eata, his abbot, to allow him to retreat completely and lay down the office of prior. He went to Inner Farne, a small rocky island about a mile offshore and seven miles south-east of Lindisfarne.

Aidan had gone there on retreat, but the island was unin-
habited. Except by demons. So that he could do battle with
them, Cuthbert built himself an oratory and a cell where
he could sleep, surrounding both with a high wall that
screened out the world and looked up only at the heavens.
His reputation for piety was growing, so even though he
had fled the world, visitors rowed to Inner Farne to seek his
blessing and his counsel. Eventually these intrusions became
something he had to control. Here is a passage from Bede:

> Then, when his zeal for perfection grew, he shut
> himself up in his hermitage, and, remote from the gaze
> of men, he learned to live a solitary life of fasting,
> prayers and vigils, rarely having conversation from
> within his cell with visitors and that only through a
> window. At first he opened this and rejoiced to see and
> be seen by the brethren with whom he spoke; but, as
> time went on, he shut even that, and opened it only
> for the sake of giving his blessing or some other defi-
> nite necessity . . . and for the sake of the sweetness of
> divine contemplation, [Cuthbert resolved] to be silent
> and hear no human speech.

Walling himself up in his hermitage, Cuthbert must have
depended on the brethren on Lindisfarne supplying him
with food and other things like firewood. But despite the
extreme austerity described by Bede and in the Anonymous
Life, the hermit did not withdraw entirely. Ecgfrith was King
of Northumbria and his sister, Aelfflaed, was Abbess of
Whitby, and she persuaded Cuthbert 'to cross the sea and
meet her at Coquet Island', a monastic community further
down the Northumbrian coast. And in a journey that sits
oddly with the life of an anchorite, whose vow is not to

leave his self-confinement, 'He went on board a ship with the brethren and came to the island' to talk to the king's sister. And their discussions were not about matters of piety and prayer. They talked about politics.

Keenly interested in the survival of her dynasty, Aelfflaed tested Cuthbert's gifts of prophecy and asked him how long her brother would live, and more importantly who would succeed him. These seem to me to be extraordinarily worldly questions for a hermit who sought only to be walled up and left in peace on a wind-blasted island – but again I may be committing the sin of separating Church and State. In both the Anonymous *Life* and Bede, Cuthbert's response to the question of Ecgfrith's longevity reads as more than a little equivocal, but on the second question he appears to have been more precise. Ecgfrith and his sister were descendants of Aethelfrith, but Cuthbert prophesied that Aldfrith, a descendant of the Deiran dynasty of Aelle, would be king. To avoid assassination, he had gone into exile on Iona, but Cuthbert seemed certain that he would take the throne of the twin kingdoms after Ecgfrith's death. And it turned out he was right.

Perhaps in return for his prophecy or to ally him and his exemplary reputation with the interests of King Ecgfrith and the descendants of Aethelfrith, Aelfflaed then told Cuthbert that her brother had it in mind to make him Bishop of Hexham. Here is Bede's fascinating account:

> Now she knew that Ecgfrith proposed to appoint Cuthbert bishop, and wishing to learn whether this proposal would be carried into effect, she began to ask him in this way: 'How the hearts of mortal men differ in their several purposes! Some rejoice in the riches they have gained, others who love riches always

lack them. You despise the glory of the world, although it is offered, and although you may attain to a bishopric, than which nothing is higher amongst mortal men, will you prefer the fastnesses of your desert place to that rank?' But he said: 'I know that I am not worthy of such a rank; nevertheless I cannot escape anywhere from the decree of the Ruler of Heaven; yet if He has determined to subject me to so great a burden, I believe that after a short time He will set me free, and perhaps after not more than two years, He will send me back to my accustomed rest and solitude. But I bid you in the name of our Lord and Saviour not to tell anyone before my death what you have heard from me!'

It is difficult to know what to make of this. No doubt it would have been awkward to refuse to meet the sister of the king, but she could have come to Cuthbert on Inner Farne. Others did. Why did Cuthbert feel compelled to re-enter the world of politics and exert himself to do so? The consequences of the meeting on Coquet Island were far-reaching and relatively immediate. In 684, King Ecgfrith did not summon Cuthbert but instead went to Inner Farne with Bishop Trumwine and a considerable retinue to persuade him to become a bishop. He had been elected at a synod held on the Northumbrian coast at Alnmouth. But when told of his elevation, he refused to take up office. This feels like another trope, a little like the traditional reluctance of popes or speakers of the House of Commons. Persuaded further, Cuthbert made some conditions. The original offer was the bishopric of Hexham, but the hermit insisted that he become Bishop of Lindisfarne and that Eata transfer to Hexham. Amidst all the rituals of reluctance and unworthiness, some hard bargaining appears to have taken place on Inner Farne.

Cuthbert will have understood that Ecgfrith wanted to associate his great reputation for piety with his kingship. He also knew perfectly well that Aldfrith was waiting in the wings and had the community founded by the saintly Columba on Iona to back him. Ecgfrith needed his own living saint. Nevertheless the contradictions were sharp. Cuthbert had long sought the solitary life of a hermit because 'he feared the love of wealth', and yet the Northumbrian church had been endowed with so many gifts of land that it had grown extremely wealthy. The hermit would become a prince of the church, and his beautiful pectoral cross made from gold and garnets is a glittering symbol of his new status. And as he walked the duneland and sheltered from the wind and rain in Cuddy's Cave, like Christ, Cuthbert had rejected 'the kingdoms of the earth'. Yet here he was, becoming one of the most powerful men in Northumbria, and indeed all Britain. On 26 March 685, Cuthbert was consecrated Bishop of Lindisfarne by Theodore, Archbishop of Canterbury, and six other bishops at York, and for two years he immersed himself in politics.

Motivation can be difficult to deduce, particularly at a distance of fourteen centuries. Perhaps in his long, shivering night vigils on Inner Farne, Cuthbert believed that God had spoken to him and directed that he take up the office of bishop. In Bede's account of the meeting with Abbess Aelfflaed, there are clearly some retrospective adjustments, particularly over matters of timing. Two years is not long to hold high office unless there was some specific task that Cuthbert wanted to achieve. But there is no evidence of that. In reality he behaved much as seventh-century prelates did, progressing around his see and all its possessions and preaching to all who gathered.

He also took up the role of royal counsellor, offering views on military matters, of all things. Against Cuthbert's advice, King Ecgfrith led an army north in May 685 to confront the Picts. At Dunnichen, near Forfar, he was defeated and killed. At the same time, Cuthbert was in Carlisle with Queen Aethelthryth, and when he had a premonition that disaster had occurred he sent the queen back to Bamburgh for her safety, behaving like a responsible, worldly counsellor.

After Dunnichen and the roll back of Northumbrian power in the north, the exiled Aldfrith became king, as Cuthbert prophesied, and a few months later he decided to lay down the office of bishop. After Christmas 686, Cuthbert returned to Inner Farne to resume the hermetic life. The contrast with his short career in politics must have been stark. It may be that, according to Bede and the Anonymous *Life*, Cuthbert was ill and believed he was nearing death. Or equally, it may be that the new king wanted a new bishop.

These two years at the centre of Northumbrian power sit uncomfortably with the rest of Cuthbert's life, but then no one behaves with unwavering consistency. Throughout my journey to Lindisfarne, I had thought a great deal about this saintly man. Had he been an icon, someone who was unbendingly perfect and remote, then I would not have found him so attractive. There are contradictions in all of us: who has not said one thing and done another?

Near the end of the Crooked Lonnen, I found the Bridge Well. Peeping out from a tangle of thick grass and withered stalks of cow parsley, it was easy to miss. The water trickled from under a capstone that looked old, as though it had been there for many centuries. Wells can be disrupted, but it is likely that Cuthbert knew this one. It used to serve St

Coombs Farm, whose name, cognate to Columba, suggests that there might have been a chapel nearby. And perhaps in the seventh century the little well was venerated. On my long walk I had met no one, and by the time I reached the village only a scattering of people walked the streets, the tide having shut a few hours before. The day was drawing to its close and lights twinkled in kitchen windows.

To watch the sun go down behind the Kyloe Hills, I climbed up the Heugh. The blood-orange ball sank quickly, its fiery light glinting off the sea. After the sun had slipped behind the hills, the edges of the cloud were gilded against the fading blue of the evening sky. For much of the day, I had been thinking about Cuthbert and his island, and without realising it I felt myself settled for the first time since I crossed the causeway. Perhaps the peace of Cuthbert had descended.

When I sat down on the stone bench, I saw two big tankers far out to sea, travelling in opposite directions. The next time I looked up, they had disappeared. Without realising it, I had spent time turning over my thoughts. On my journey to Lindisfarne, it had occurred to me more than once that my efforts to concentrate on issues were mostly fleeting. I am easily distracted. But up on the Heugh I did manage to focus for once. Perhaps the island and its silence were working on me. Though in truth my feelings were more negative than positive.

Nearing seventy, I am past caring what other people think of me, except for family and close friends. I have come to a simple, even simplistic view that what I think of myself and my actions is what matters.

On my way to supper, pondering my advancing years, thinking that the years raced past faster and faster, I remembered an old friend's analysis of the stages of later life. Ricky

Demarco built a reputation in the 1960s, '70s and '80s as a gallery director and gifted cultural entrepreneur in Edinburgh, and the last time I saw him he told me he was eighty. Using a football analogy, surprising for Ricky, he said that we might all reasonably expect to fulfil the biblical three score and ten, but anything after seventy was extra time, and after eighty it was the penalty shoot out.

When God Walked in the Garden

I woke to wind-driven rain being flung against the window-panes, and immediately thought of our old horses. Four of them, the Old Boys, live out through all weathers, well-rugged and fed, but wind-driven rain is a bane for them since it is so penetrative. By pointing their hindquarters into the direction of the wind, sometimes standing close to each other and where the undulations of the East Meadow take the edge off the cold blast, they try to weather the worst of it. All are big horses who were ridden by my daughters and my wife to high competitive standards, but now they are in their twenties. Gem, a cross-bred Dales pony, will be thirty this year. In the terrible snows and rains of 2017, we lost Murphy, a lovely, big, elegant chestnut I sometimes rode. Slipping on a slope in the meadow, he went down and could not get up again, his rickety, arthritic legs having no purchase in the mud. When the vet came to put him to sleep (I dislike euphemism but cannot bear to write anything else here), Lindsay gave him a bowl of hard feed and he amazed us by finishing it down to the last crumb. His will to live was not extinguished, but we had no alternative and, in what my wife calls the last act of kindness, he died a peaceful death. We covered him with old rugs and weighted them with stones to keep the foxes off him overnight, before the mechanical digger came to bury him.

I think often of the horses and the other animals who have died. Tears sometimes come, but I know they had good lives and we loved them. There should be no regret in that. They were lucky to be with us and we were lucky to have them.

As the wind whistled and the rain spattered, it was no day for walking – although Cuthbert and the more austere brothers may have seen it as another God-given opportunity to mortify the flesh. Their thick, homespun woollen robes will not have kept out rain like this, no matter how much lanolin had been left in the yarn. Once they were wet through, they would not have been dry for days. No need for hair shirts on Wild Lindisfarne.

Instead, it was a day for a book. Not the sort to be read in an armchair in front of a crackling fire, but one of the most famous books in the world.

The *Lindisfarne Gospels* were written and painted on the island, a blaze of glorious colour against the background of the grey North Sea. There is a facsimile copy in St Mary's Church and I had brought with me an excellent commentary, *Painted Labyrinth* by Michelle P. Brown. Sadly, the real thing, the *Gospels* themselves, are in the British Museum in London, something that strikes me as a national disgrace and a waste. Objects acquire much more meaning if they are seen in their place of origin rather than as one of a thousand treasures in a vast museum that has no connection with them. If they were shown where they should be, in a bespoke building dedicated only to them on Lindisfarne, the presence of the *Gospels* would be transformative, bringing visitors all year round, even in weather like today's, to look at this stunning work of art in precisely the place where it was created.

Visitor numbers for the British Museum would suffer no ill effect – there is so much else to see – but the island economy would prosper. And more than that, this remarkable book would be better understood and its artistry relished all the more in the place where its creators walked their lives thirteen centuries ago.

The facsimile is very good, a gift to the church, and it sits close to the monochrome wooden sculpture of the monks carrying Cuthbert's coffin. That is pleasing because it was the power of the saint who prompted the creation of this work of art. The *Gospels* were written and painted by one man, probably over a two-year period some time after 698, the year Cuthbert's body was found to be uncorrupted and he was elevated to sainthood. The Bishop of Lindisfarne, Eadfrith, is credited with authorship in a colophon appended to the *Gospels* in the tenth century by a priest called Aldred:

> Eadfrith, Bishop of the Lindisfarne Church, originally wrote this book, for God and for Saint Cuthbert and – jointly – for all the saints whose relics are in the Island. And Ethelwald, Bishop of the Lindisfarne Islanders, impressed it on the outside and covered it – as well he knew how to do. And Billfrith, the anchorite, forged the ornaments which are on it on the outside and adorned it with gold and with gems and also with gilded-over silver – pure metal. And Aldred, unworthy and most miserable priest, glossed it in English between the lines with the help of God and Saint Cuthbert . . .

Aldred translated the Latin gospels into Old English at Chester-le-Street, where the Lindisfarne community built a church in the tenth century before moving on to Durham. Aldred also noted that the book was bound by Ethelwald, who succeeded Eadfrith, and he makes the point that he 'well knew how to do [it]'. Clearly this was not the first illuminated manuscript either man made, such is the richness of their skills and experience. But it is the only one that survived. The skill of Billfrith has been lost, probably looted in the Viking raids on Lindisfarne. Surprisingly, he was an anchorite, a hermit who had himself walled up, although clearly not permanently. Metalworking such as he did needs practice and also the *Gospels* are unlikely to have been his only work. It seems a strange combination – anchoresis and the skills of a jeweller – and perhaps it should qualify our views of these ascetics. Their privations are unlikely to have been constant and were probably periodic, perhaps like going on retreat.

The *Gospels* are a stunning artistic and spiritual achievement that show an apparently small and marginal island community at the centre of Western European cultural life. For the main text, Eadfrith used a southern Italian gospel book and consulted other works copied around the shores of the Mediterranean. Celtic and German styles of metalworking are imitated in the illumination, and the carpet pages that come between each gospel are like oriental rugs. But even more than all of these connections, the *Lindisfarne Gospels* are remarkable because they are clearly the product of a community with the skills and knowledge to translate all of these into one of the greatest books ever made. There must have been a substantial library on the island, a store of plants identified and cultivated to create the colours and make the ink, and a huge investment in calfskin for the pages. In every sense the *Gospels* are a spectacular outpouring from a deep reservoir of immense and varied skills and knowledge, something that may be seen as unsuspected on this windblown little island off the northern edge of England.

That is a further and persuasive reason why the book should come home: it would change perceptions, both outside and inside.

The four gospels of Matthew, Mark, Luke and John each begin with a preface and what are known as Canon Tables. These list passages of text and show which are shared in which gospels. Framed by architectural arcades held up by richly decorated columns, they are beautifully written, the script an artform in itself. Then come the incipits, literally 'it begins', the opening words of the gospels, and these are gorgeous, the first letter made into a virtuoso piece of decoration that takes over the whole page. Each of the four evangelists is portrayed, all of them as scribes, composing their gospels. Their iconography is distinctive. Matthew is shown as an older, grey-bearded man and behind his head there is a figure that might be an angel, blowing a trumpet, possibly the Last Trump on the Day of Judgement. Looking out from behind a beautifully painted red curtain is a man carrying a book, a figure that may represent Christ. Mark is shown with a lion (hence the lions on St Mark's in Venice), and he is beardless and youthful. So is John, and behind him, carrying a book, an eagle flies directly to the throne of God, prefiguring the Second Coming. Luke is shown as older and bearded, and behind his halo is a winged calf, a symbol of the crucifixion. In all the portraits, the drapery is luxuriant, wonderfully well realised by Eadfrith, an artist at the peak of his powers.

These portraits are the sole representations of people in the *Gospels*. Early Christians had a fear of idolatry, the graven images of the Bible, and this meant that abstract, geometric decoration of the sort seen on the carpet pages and the calligraphy in the book were given so much prominence. The *Lindisfarne Gospels* are emphatically not a riot of gorgeous colour, as I read in one history, but a contained, controlled piece of work that uses the harmony of proportion and the precision of intricate design.

One hundred and thirty calfskins were used, and no doubt more were discarded because of blemishes. The Anglo-Saxons were fond of riddles, and in St Mary's Church a display board reproduces one that captures the process of preparing the skins to be made into pages:

An enemy ended my life, deprived me of my physical strength. Then he dipped me in water and drew me out again, and put me in the sun where I soon shed all my hair. After that, the knife's sharp edge bit into me and all my blemishes were scraped away. Fingers folded me and the bird's feather moved over my brown surface, sprinkling meaningful marks. It swallowed more wood dye and again travelled over me leaving black tracks. Then a man bound me, he stretched skin over me and adorned me with gold. Thus I am enriched by the wondrous work of smiths, wound about with shining metal.

The word 'vellum' comes from *velin*, an old French word for a calf, and traces of the animals can still be seen in the *Gospels*. The spine left a darker line and the sheets were arranged so that it runs horizontally. Apparently, this prevented the cut calfskin from cockling, trying to return to the shape of the animal. This would have caused paint to peel and flake. Incidentally the shape of the calf itself meant that skins were cut as double pages, written and painted out of sequence and then bound in gatherings of even numbers, eight pages in the *Gospels*. This established the shape of modern printed books.

Once the skins had been soaked, scraped and cut to size, Eadfrith ruled the pages with a stylus that would leave an impression rather than a mark. He then pinned the skin to a writing board. Good light was at a premium and on bright, windless days scribes worked outside rather than burn candles in darker interiors. Using a penknife, Eadfrith then cut his quill pens (from geese, swans or even crows – bird life on Lindisfarne around the year 700 was probably as plentiful as it is now) and made sure each nib was of a uniform breadth so that there was no variation in the size of his script. And he succeeded. The lettering never appears to vary and its uniformity makes it look as though it might have been printed. Ink was made from oak galls, the wood dye of the riddle, mixed with iron salts. Sometimes known as oak apples, these are growths on the tree that contain the larvae of a species of wasp. This must have been easily available, but probably not on treeless Lindisfarne.

Once all was prepared, Eadfrith began to copy from his southern Italian gospel in a script known as insular majuscule, sometimes as half-uncial. It was first developed in Ireland and came to Lindisfarne with Aidan and the early bishops from Iona. Majuscule refers to the size of the lettering, and it is indeed majestic, flowing across the pages. It is very time-consuming to write and the scale of Eadfrith's achievement is humbling. But it is a difficult text for modern readers, even those with a little monastic Latin. There is no punctuation, but the length of the line usually helps clarify meaning. When a sentence ends, the rest of the line is left blank.

When he ruled his incipit pages, Eadfrith had already worked out the decorative scheme and allowed space for it. Analysis of the *Gospel* pages show something surprising. Once he had tried out his ideas for decoration on some waste material, Eadfrith used the back of the sheets of calf-skin for the outline drawings he needed to guide his painting. That meant he drew them in a mirror image, and in order to see the marks through the skin, he appears to have invented the lightbox. Candles were placed behind the skin, which must have been held rigid on some sort of stretcher, or if working outside, polished metal was angled to reflect the sun so that he could make out his marks. And to avoid scoring the sheets, making marks that might have trapped paint, he used a more rounded lead point, something he seems also to have invented.

The colours are glorious and all seem to have been made from plant or mineral extracts on Lindisfarne. Other illumin-ated manuscripts used pigments from around the world – red or vermillion came from insects that lived under the bark of oak trees that grew around the Mediterranean, while the blue of lapis lazuli came from the Himalayas. All of the colours used in the *Gospels* appear to have been home-made. The group who helped Eadfrith were clearly skilled in plant lore, extracting reds, greens and yellows from what grew locally. It was as though the colours of the island found their way onto Eadfrith's pages, as though a garden made by God was used to praise Him. Not all of the pigments came from plants; other parts of Creation also added vivid hues. Yellow came from orpiment, a type of mineral arsenic that was extracted from ore. Purples, crimsons and blues could be made when acid or alkaline materials were mixed with plant extracts such as lichen, folium and woad indigo. These recipes, particularly for the ink, were excellent because over thirteen centuries the colours have faded only a little.

Above all, Christianity was and continues to be a religion of the book and the word. The Old and New Testaments were expressions of the revealed word of God and also clear guides on how to live a pious life. As the cult of St Cuthbert grew stronger, Eadfrith realised that it needed a gospel book as a focus. In the 670s the supporters of the rival cult of St Wilfred at Ripon commissioned a gospel written in gold on purple paper, somewhat in the extravagant style of the saint himself. There is also a sense of Eadfrith's work itself as something miraculous, touched by the divine and an object for reverence. Much later, the *Lindisfarne Gospels* were chained to the high altar in Durham Cathedral and into the pages of the holy book records of gifts given to the prince bishops and their see were inserted, perhaps to sanctify them. The colour and artistry of Eadfrith's work must have seemed astonishing to contemporaries, but the fact that this was the work of a man was made clear by the deliberately uncorrected mistakes and parts that were left incomplete. Only God was capable of a work of complete perfection.

*   *   *

High winds have only one welcome aspect: they drive the clouds quickly, and by midday the rain had stopped. Day trippers were emerging full of cake and coffee from overcrowded cafes. On such a foul morning, I was surprised that so many had crossed the causeway. Seeing the crowds in the streets, I realised that the solitary young pilgrims I had noticed were exceptional in another way. The vast majority of visitors were old, older than me, probably many of them retired and free to come to Lindisfarne on a weekday. Some were very elderly indeed, and I stopped counting the number who could not walk unaided, either with an arm through another or a Zimmer frame with wheels. But they did it. They came, even though it needed great effort, and I admired that. Lindisfarne Leg was nothing compared to some of the disabilities I saw. Many may have been doing something similar to me, coming to Lindisfarne to face personal demons, take stock of their lives and ultimately learn how to die, how to find what they needed to make a good death.

By the road running around the Ouse, the small bay between the village and the castle that acts as a natural harbour for fishing boats and other craft, there is a new building that from the outside seemed to have no obvious purpose. Single storey and beautifully stone-built, it turned out to be a large version of a hide, a place where the birds that land on the southern fields of St Coombs Farm and the pond close to the road might be watched. Inside are large prints of photographs of the island and a vast window that looks north towards where birds might be. It was a peaceful place and empty when I entered, even though many were passing on their way to and from the castle. Several of the photographs have quotes from visitors printed in the corners, people who sound like the pilgrims I had briefly spoken to.

When I come back over the causeway, it's like a portal. I'm going back to peace, tranquillity and sanity – away from the mania of England.

The sound that most visitors notice here is the silence.

If you stop and just listen, wherever you are on the island, you can hear the sea.

I remember lying in the grass listening to skylarks and everything felt right. Now when I hear them, it's as if it was then; time has shortened and I go back. Even though lots has changed here, some things haven't.

The tide was not yet shut and I decided to visit the castle while it was still open. It dominates the island, its walls seeming to grow out of the dolerite rock it sits on. Its commanding exterior has featured often in film and on television. When I travel south on the London train, my spirits sinking, it is a landmark I always seek from the window as we hurtle past. For the last year or two, I noticed the castle had been shrouded in scaffolding and white sheeting, but when I arrived on the island I was glad to see it revealed once more, the restoration project having been completed.

In a lovely shop fitted inside the hull of an upturned boat (I later discovered that it was one used in the dangerous and clandestine North Sea traffic of the Second World War with occupied Norway) I bought my ticket and sought out one of the English Heritage guides. They are often very knowledgeable and I wanted to know if the restoration work, which had pared back a substantial part of the fabric of the castle, had discovered any traces of previous fortresses on the rock. But I was disappointed. When I asked the young man what was there during Cuthbert's time, he replied, 'Nothing. There was nothing here. This was the windy end of the island and nobody wanted to live here.' I thought that highly unlikely. Throughout much of history there have been more compelling reasons than the weather for choosing to build on a site. In addition, the rock was known as Beblowe's Crag, a name cognate to Bebbanburh or Bamburgh, so called after Aethelfrith's queen. The crag on Lindisfarne would also have been seen as a superb site for a fortification, easy to defend and with a commanding all-round view of the sea and the island. But of course nothing will now remain of its wooden palisade and any artefacts will be irretrievably lost under the later fabric.

The castle first comes securely on record in 1548–9, but, close to the volatile border with Scotland, it is likely that the crag was fortified during the medieval period. The community of monks had been scattered after Henry VIII ordered it to be dissolved in 1537 and stone was carted around the bay from the deserted priory to build a fortress. Its cannon would protect the harbour. England and Scotland had been at war for much of the first half of the sixteenth century and Lindisfarne had become a strategically important naval base. In 1543 the island was heavily garrisoned and warships lay at anchor in the Ouse. To guard the western end of the natural harbour, three earth and timber forts were thrown up, and in 1670 a stone fort known as Osborne's Fort was built. In those days the harbour reached farther inland, close to the eastern walls of the priory and as far as the foot of the Marygate. The Union of the Crowns of 1603 and the Parliaments in 1707 made most of these defences redundant. By 1820 the castle had become a coastguard station, but by the end of the nineteenth century it had been abandoned and the weather was destroying the fabric.

Edward Hudson had made a great deal of money from *Country Life*, a magazine he founded, and in 1902 he bought the much dilapidated Lindisfarne Castle. Sir Edwin Lutyens designed many country houses, the Cenotaph in London, was instrumental in the planning and architecture of New Delhi and had been the architect of the head office of *Country Life*. Hudson commissioned him to rebuild the castle and, after nine years, the exterior was beautifully realised, especially the western and southern facades. And work also went on off the site. From the northern windows, a new garden could be seen. Walled on three sides, it was designed and planted by Hudson and Lutyens' friend Gertrude Jekyll, and it is a miracle of persistence. Having to take into account frequently hostile weather, she succeeded in creating a garden with a simple shape and planted in it many brilliantly and subtly colourful – and hardy – flowers and shrubs. Her scheme to have alpine and other varieties planted directly into the castle rock itself was dramatic. A boy was lowered down from the walls in a basket and into the crevices he shoved earth and bedded in young plants. Very few survived the salty winds and rain.

The young guide I spoke to said he would be giving a talk in the Ship Room, but I had time to look around the interior of the castle first. All of the rooms had been emptied of their furniture during the two-year restoration and in its place there was an exhibition of modern art.

Even though they looked stark and cold without furniture, the interiors of the castle were a surprising contrast with the drama, completeness and warmth of the exterior. They seemed frankly dismal, poky and badly lit, with tiny windows. Winter supplies enough darkness in my life, and I like interiors to be bright and full of all the different sorts of light our climate gives us.

I sat down on a stone window seat in the Ship Room to listen to the young guide's talk and he seemed nervous. Running quickly through the early history of the castle, he came to the period of Hudson and Lutyens. There was one interesting nugget I didn't know. Lutyens had thought the original tall chimneys were ugly and reduced them by half. This meant that the open fires did not draw properly, and the guide pointed out how blackened with smoke the stone mantle in the Ship Room was. It occurred to me that smoke probably also leaked into the rooms themselves, to say nothing of clothing, furniture fabric and curtains. Something of a design fault, Sir Edwin. As I was deciding Lindisfarne Castle was not a place I would like to live, another guide rushed in to say that because of the wind, the tide would shut earlier than the advertised times. And because, once again, the staff lived on the mainland, the castle would also shut.

As the visitors fled to their cars, I walked down to watch the tide refloat the boats anchored in the Ouse. Fortified earlier by an excellent crab salad with extra chips at the Ship Inn, I sat down on the bench with the cable-tied messages under it. The long views south to Bamburgh and the rocks of the Farne Islands are very soothing, little changed from when Cuthbert walked around the bay, and in the warming sun and the brisk wind I felt settled.

I came to the Ouse because I wanted to think about Lindisfarne's long past and a turning moment in Britain's history, something that took place precisely where I sat. I have often found that walking where Roman soldiers marched or being in old churches where pilgrims came can make time collapse on itself. Sitting on the edge of the Ouse, terrifying wraiths raced past me, with murder and plunder in their ghastly hearts. On 8 June 793, what medieval chroniclers called 'a shower of hell' burst over Lindisfarne. The entry in the *Anglo-Saxon Chronicle* wrote of fell portents:

> This year came dreadful forewarnings over the land of the Northumbrians, terrifying the people most woefully: these were immense sheets of lightning rushing through the air, and whirlwinds, and fiery dragons flying across the firmament. These tremendous tokens were soon followed by a great famine: and not long after . . . the harrowing inroads of heathen men made lamentable havoc in the church of God in Holy Island by rapine and slaughter.

The Vikings had sailed into history. Blown westwards across the North Sea by greed and a need for adventure, their dragon-ships, the *dreki*, were seen by the monks off the island coast. They looked different from the ships that usually plied the coasts. Roared on by their sea-lords, the Viking oarsmen rowed hard for the stony beach to rasp up their *dreki* above the tide-line. And as the monks scattered, these pagan worshippers of Odin, Thor, Freya and Tiw did not hesitate to break down the doors of the church and steal everything of value that they saw. Any monks who stood in their way were cut down and others no doubt fled to the dunes. It may be that Billfrith's jewel-encrusted cover for the *Lindisfarne Gospels* was ripped off the great book in a Viking raid. It is a miracle that the *Gospels* themselves survived. In the priory museum, what is known as the Domesday Stone shows warriors attacking, their swords and axes raised above their heads ready to strike. It may be part of a memorial to those whose blood spilled over the shrine and Cuthbert's coffin on that terrible day.

Shockwaves reverberated throughout Western Europe. The monastery and its treasures were undefended; there had been no need. According to Bede, Lindisfarne was 'the very place where the Christian religion began in our nation', and God and the saints protected it. When news of the raid and its devastation reached Alcuin of York, a scholar who had gone to the court of Charlemagne and was considered by contemporaries 'the most learned man anywhere', he was stunned. His letters to Higbald, Bishop of Lindisfarne, have survived and they reflect shock, and recrimination. 'The church of St Cuthbert is spattered with the blood of priests of God, stripped of all its ornament, exposed to the plundering of pagans,' he wrote. More than shocked, Alcuin could not understand why God and the saints had allowed the Vikings to pillage the monastery. There had to be a reason. Here, he asks Higbald to examine his conscience:

Either this is the beginning of greater tribulation, or else the sins of the inhabitants have called it upon them. Truly it has not happened by chance, but it is a sign that it was well merited by someone. But now, you who are left, stand manfully, fight bravely, defend the camp of God.

Alcuin may have had a point. Chroniclers recorded a conspiracy in 788 against King Aelfwald of Northumbria led by a nobleman called Sicga. The king was killed but the *coup d'état* appears not to have succeeded. Faction fighting continued, and five years later Sicga 'perished by his own hand'. But on 23 April 793, only a few weeks before the Viking attack, this regicide and suicide was buried on the holy ground of Lindisfarne, presumably because his family had given gifts to the church. The incident appears to have been well known, the occasion of disapproving comment, and it may be that when Alcuin wrote to Higbald that the attack was merited, he might have had Sicga in mind.

The summer after the raid on Lindisfarne, Viking dragon-ships were seen in the Hebrides and the diseart founded by St Donan on the island of Eigg was raided. In 795 Iona was attacked and many of the treasures of St Columba were stolen. The Vikings returned in 798 and 802, and in 806 they slaughtered sixty-eight monks and lay brothers. Abbot Cellach had no option but to make plans to abandon the island. On Lindisfarne, there are no records of more raids; perhaps warriors from Bamburgh garrisoned the island, using a stockade on the castle rock. Projecting into the sea, it was an excellent place to keep lookout for dragon-ships, and where a beacon could be lit to raise the alarm and rally resistance if an approach was made through the harbour. Used to undefended monasteries and easy pickings, the Vikings may have thought it too much trouble.

Raids around the coastal communities continued throughout the early decades of the ninth century, as fleets began to cross the North Sea. Some established *longphortan*, ship-camps where they had dragged the *dreki* inland or upriver and built a stockade so that they could overwinter and resume raiding in the spring. It was the beginning of colonisation. One of the earliest ship-camps was set up at the mouth of the River Liffey, what became Dublin, and it was there the Vikings established their most lucrative activity. The early attackers stole portable loot, whatever glittered on the altar tables of monasteries, but they also took whoever could be abducted, bound and bundled onto the open deck of a dragon-ship. From the first, the Vikings took captives, and in one of his five letters about the raid on Lindisfarne, Alcuin offered diplomatic help in ransoming 'the youths who had been led into captivity'. At Dublin a busy slave market was established and, rather than jewelled crosses or gospel book covers, it was human trafficking that made Viking sea-lords wealthy.

The conventional view of these raiders as indiscriminate butchers was inaccurate, a shocked reaction to the early raids and particularly the slaughter on Iona in 806. Monastic scribes called them the Sons of Death and, like a bolt of Thor's lightning, they seemed to appear out of nowhere, sailing through the sea-mists and trailing devastation in their wake. In reality the Vikings wanted captives not corpses. Merchants sailed from Muslim Spain and the North African coast to buy Christian slaves at the Dublin slave market. Fair-headed men and women were sold at a premium and sometimes customers requested that they be castrated. Aristocratic captives also fetched high prices. Evidence from an analysis of ancestral DNA shows a scatter of Scottish markers in the modern population of the western seaboard of Norway. This may be a legacy of the slave trade, the descendants of captives brought back across the North Sea.

For the year 841, the *Anglo-Saxon Chronicle* reported a resumption of raiding down the East Anglian coast and the old kingdom of Lindsey, now north Lincolnshire. Bishop Ecgred of Lindisfarne may have felt uncomfortable at how close the threat of attack was creeping and anxious that the sails of a fleet might be seen offshore, rather than two or three ships. With his prior and chapter, he made plans to abandon the island. The traditional date for the departure of the monks is 875, but there is evidence that it took place thirty years before.

In 655 King Oswy of Northumbria had gifted land to Lindisfarne at Ubbanford on the lower reaches of the Tweed. This place became known as Norham, originally the Northern Settlement, and it watched over the first ford upriver from Berwick. A wooden church was established and, soon after he became bishop, Ecgred had a stone church built on its footprint and dedicated to St Ceolwulf. It has disappeared, its site thought to be under a stand of yew trees to the east of the large medieval church. Around it, there is a good deal of evidence to suggest that something more than a church was built there during Ecgred's time. The very large churchyard hints at a monastic precinct, and on the village green there is a preaching cross, possibly a successor to something older. In the medieval church, there are individual fragments of cross shafts. Clumsily cemented together in a column are many other fragments, including a carving of an angel, that came from other crosses around an earlier church. For more than a century, Norham had also been the mother church, the *matrix ecclesia*, as well as the administrative centre of Norhamshire, an extensive and early royal grant to Lindisfarne.

As the threat of Viking attack edged closer, Bishop Ecgred led a sad exodus across the sands and the Holy Island of Lindisfarne was abandoned. The wanderings of the Congregation of St Cuthbert had begun. All of his relics and those of other saints, such as Oswald and Aidan, were loaded onto carts. Cuthbert's coffin may indeed have been carried by relays of bearers, as suggested by the sculpture in St Mary's. It is also thought that Aidan's original wooden church, itself a sacred relic, was dismantled and brought to Norham. It must have been a time of immense sadness, to leave the place where saints had walked, the island-church built by God and recognised by Aidan. But Cuthbert's body and all of the holy relics had at all costs to be preserved and kept out of the desecrating hands of heathens. It may have been in the years after 841 when the ditches and banks of a precinct were dug at Norham and crosses raised and painted. Perhaps the pillar in the medieval parish church (dedicated to Cuthbert, St Peter and Ceolwulf) contains fragments of crosses that originally stood on Lindisfarne. If so, they are a sorry sight, but at least they are preserved.

It was to be a temporary place of shelter. Led by Halfdan, part of the Great Heathen Army invaded Northumbria, and in 875 the Congregation of St Cuthbert loaded their carts once more. Here is an extract from the Chronicle of Symeon of Durham:

> They wandered throughout the whole district of Northumbria, having no settled dwelling-place; and they were like sheep flying before the face of wolves . . .

Many legends swirled around the wanderings of Cuthbert, his holy relics and the community who carried them. According to Symeon of Durham, the monks took ship for

Ireland, the *Gospels* were washed overboard, Cuthbert appeared in a vision to Hunred, one of the bearers of the coffin, and told him that the great book had been washed up on a beach near Whithorn, the shrine of St Ninian, its colours miraculously intact. Analogies and borrowings from biblical stories from the Old Testament such as the exodus from Egypt, the search for the Promised Land, the Babylonian Captivity were recognised by contemporaries and the prestige of the cult of Cuthbert gained much. The perils of the Irish Sea may have had a part in the story, but it seems likely that the wanderings were more restricted in scope and limited to the north of England.

What surprises me, miracles and divine intervention aside, is how the Congregation of St Cuthbert survived in territory controlled by Halfdan and his Great Heathen Army. In fact, they prospered. Symeon of Durham noted that the monks were given rich gifts and sometimes granted land as they wandered. Perhaps they were sufficiently wealthy to employ soldiers to guard them and their precious luggage. Prelates with large retinues were by no means unusual and were criticised by Alcuin in his letters. Bishops should comport themselves with humility and follow the examples of the Bible, he wrote, but in reality most were noblemen and women and they behaved much like their secular relatives. An alternative hypothesis might suggest that such a tempting target for Halfdan's warriors must have enjoyed his protection. Perhaps in his efforts to establish control over the old kingdom of Northumbria, he needed allies and some unity, and amongst the people Cuthbert's cult was clearly becoming powerful.

When the Congregation reached Crayke in Yorkshire, close to York, the focus of the Viking kingdom, a remarkable and eloquent episode played out. The monks became involved in a political coup. After the death of Halfdan in 882, Cuthbert appeared in a vision to Bishop Eardulf and told him to find and ransom Guthfrith, a Danish Christian, and support his bid to become king of Northumbria. And an unlikely chain of events took place, as the Congregation succeeded in having their candidate crowned. It was an episode that showed them not as a band of homeless, fearful pilgrims pushed from pillar to post but as power-brokers. Guthfrith reigned between 882 and 895, gave grants of vast tracts of land to the monks, and most importantly made it possible for them to find a new home. Inside the welcome fortifications of the old Roman fort at Chester-le-Street in County Durham a splendid church was built, and it was there that Aldred translated the *Lindisfarne Gospels* into Old English and added his colophon.

For a century, the community seemed settled inside the fort, but in the 990s the *Anglo-Saxon Chronicle* reported an intense period of Scandinavian raiding. Bamburgh was destroyed in 993 and the settlements around the mouth of the Humber were attacked. Two years later the monks' carts trundled out of the gates of the fort, bound for the monastery at Ripon in West Yorkshire. Far inland, close to the mouth of Nidderdale and the foothills of the Pennines, it probably seemed a safe haven. But after only four months the Congregation abruptly decided to return to Chester-le-Street. On the way back north, they halted. In a baffling twist in the tale, legend insists that the monks had followed two milkmaids who were looking for their dun cow. Why? What was a long caravan of precious baggage and the monks who looked after it doing following milkmaids? Perhaps they were attractive. Anyway, they led the Congregation to the banks of a river. The cart carrying Cuthbert's shrine stuck in the mud and refused to move. It seemed that the saint had decided he need go no further. The Congregation had followed the milkmaids to a loop in the River Wear. Just as at Old Melrose, the river had created a peninsula with water on three sides and only a narrow neck of land where it turned back on itself. But it was not its attraction as a diseart that persuaded the monks to stop. The site was elevated, with high river cliffs to the south, and so was much more easy to defend than Old Melrose. Cuthbert had made a good choice and, after three days of prayer and fasting, the monks agreed with him. They began to build a wooden church on the site of what became Durham Cathedral.

Bishop Aldhun was well connected locally with Northumbrian nobility and could rely on protection as well as gifts of money and land. The wooden church was quickly replaced by the White Church of Stone, and after it was consecrated in 998 Cuthbert's coffin and his shrine had at last found a home. And the monks had found an income. The White Church was completed in 1018 and one of the earliest of many pilgrims who came to the river peninsula at Durham was King Cnut. Ruling over what amounted to an empire of the North Sea that encompassed not only England but also Denmark and Norway, this immensely powerful and capable man both endorsed the cult of Cuthbert and was endorsed by it. Pilgrims began to come to Durham in large numbers and the bishopric became extremely wealthy and widely landed.

In 1080, William the Conqueror recognised that the see of Durham was strategically vital, as he consolidated his hold on his new kingdom. Needing a bulwark in the north against the Scots, he appointed William de Carilef as the first Bishop Palatine, what became known as a prince-bishop. In addition to governing the church's lands up to and beyond the Tweed, the Bishop Palatine was given extensive secular powers, both military and civil. With the right to raise taxes, try all legal cases without any exceptions and raise troops, he could act independently as though he was king in the north. Speed of reaction to invasion was the reason why the king in the south created the Bishopric Palatine to hold the north. At that time, the building of the present cathedral began in earnest and a great deal of money was needed. When Ranulf Flambard, a successful civil servant and courtier, was made bishop, he paid little attention to spiritual duties. One monkish chronicler sniffed at his morals, writing that he liked to have his lavish meals served by ladies who wore tight-fitting bodices. At Norham, Flambard spent his time not at the church where the community had worshipped but in a mighty castle he had built on an elevated site above the Tweed. It glowered over the river at Scotland. Cuthbert had fled to the secret tracts of solitude because he feared the love of fame, and of wealth, and yet his sanctity had produced such great wealth that his successors became tremendously powerful magnates who did not hesitate to carry his standard into battle.

Lindisfarne had been completely deserted after the departure of Bishop Ecgred and the monks, and for 200 years the island disappeared from the historical record. When the Congregation carried Cuthbert's coffin across the causeway, it seems that no one was left behind. In the monastic precinct the grass grew tall and the wind whistled around the chapel on the Heugh. The place of saints and spirits had become a place of ghosts, the church made by Creation and the paths of righteousness through the dunes were deserted, a wilderness, a broken world abandoned to the birds, the seals and the creatures of the land. Only the dead remained – Beannah, Osgyth, Ethelhard and the others – their bones buried forever in holy ground.

Eleventh-century politics shone a sudden spotlight on Lindisfarne. In 1069–70, the north rebelled against William the Conqueror and the magnates on either side of the Pennines were supported by a huge Danish fleet. When the king marched north to crush the rebels, Aethelwine, the last Anglian bishop of Durham, fled, taking the relics of Cuthbert with him. Once again they crossed the sands to Lindisfarne and a twelfth-century manuscript shows two monks carrying an elaborate shrine. They did not stay long. By 1072 Aethelwine had died in a dungeon and a Fleming, Walcher, had been appointed bishop in his place. He was safely loyal to William I. And so, for a thousand years, Cuthbert's bones have found their rest inside the magnificence of Durham Cathedral.

*   *   *

After a foul start, the day had cleared completely and, even after the sun had slipped behind the Kyloe Hills, some late evening light warmed the village. The clear sky would mean a shining, bright full moon and after supper I walked down to the harbour to watch its beams trace a pale, rippling path across the surface of the sea. It was my last night on the island. Instead of going to bed, it seemed right to walk. Beyond the street lights of the village, there was enough moonlight to see my path and I started up the lonnen that leads from Chare Ends to the dunes. The winds of the morning had died away almost completely and when I climbed up to the first range of dunes, there was little more than a light breeze, a zephyr. Turning to look back at the lights of the village, and of Bamburgh Castle and the Farne lighthouse beyond, I sat down on a sandy shelf on top of the high dune. It was what I had seen in 1965 when I was little more than a child, a fifteen-year-old boy looking out over what life might hold. The years between seemed heavy, suddenly, and I felt tears come, sadness that all that time was behind me and so little remained. I thought about my mum, my dad and Bina, my grannie, who looked after me when I was a wee boy needing a cuddle. I still miss them, and up on the sand dune I wept because of something simple and incontrovertible. I had lost them, these people who made me.

Their departures began with the death of my dad on a snowy February night in 1986. I was working in my office at home when the phone rang. The doctor told me that my dad had died, apparently from a heart attack, about an hour before. He had been standing by the sitting-room window, had put his hand to his chest and said to my mum, 'Oh, Ellen,' and fell back against a chair, said the doctor.

'Tell my mum I will be there with her as soon as I can.' It was snowing steadily, big flakes, and I loaded the back of my car with anything heavy I could find to keep it down on the road and make it as stable as possible. Soutra would be treacherous, the hill the A68 from Edinburgh has to cross before the road descends to Lauderdale and the Tweed Valley. I phoned the police in Lauder to say I would be crossing it and they agreed to look out for me. At Fala Village, I stopped at another phone box to call them, so that they knew roughly when I would cross Soutra. It was snowing so heavily that the windscreen wipers were almost useless and often I had to guess where the road was in the whiteout. But I made it, keeping focused by playing music very loud, driving everything out of my head except the thought that I must be with my mum. On this night, no matter how bad the weather, she could not be alone.

When at last I stopped outside 42 Inchmead Drive in Kelso, Margaret Boyd, our kind and gentle neighbour, met me at the front door. My mum was hunched by the fire, staring at it, in shock, and when I embraced her she felt limp, exhausted. Adam, Margaret's husband, was an immensely strong man and he had carried my dad's body upstairs to the bedroom. When I went up, I lay next to him for a while, holding his cold hand. After a time, Margaret knocked on the door. 'Alistair, your engine is still running.' When I went out to turn the key, I looked down the road, its yellow streetlights only just visible through the falling snow, and I thought that one day my son would make this journey.

I stayed on the dune for a long time, watching the moon move slowly across the sky, thinking about the dead and about my children and grandchildren and their lives to come. Cuthbert had a powerful faith and no doubt it was much tested, especially towards the end of his life. What sustained him was an absolute belief in an afterlife and the bargain he had made with God. In exchange for all of that prayer and privation, Cuthbert would ascend to heaven as Aidan did, carried aloft in the arms of angels. I have no transaction like that, nothing to offer and no belief in a life after death. And so I simply have to be accepting of the inevitable and try hard not to be afraid. As I walked back down the lonnen, I hoped I would be lucky and thought of one of my heroes, Robert Louis Stevenson.

As a child and a teenager, I loved *Kidnapped* and *Treasure Island* and still believe him to have been a very great writer. He was at his peak in the 1880s with these novels, and *Jekyll and Hyde*, *The Master of Ballantrae*, *The Black Arrow*, *The Wrong Box* and the wonderful *A Child's Garden of Verses*. But his health began to deteriorate and he often moved in search of a climate that would ease his respiratory problems. *Kidnapped* was written in Bournemouth. Eventually he settled on the Pacific island of Samoa, but became more and more depressed.

His powers seemed to be fading and he feared a decline into permanent ill health: 'I wish to die in my boots; no more Land of Counterpane for me. To be drowned, to be shot, to be thrown from a horse – aye, to be hanged, rather than pass again through that slow dissolution.'

Then he began work on a new novel and energy returned as he wrote *Weir of Hermiston*, a work he would not complete. 'It's so good that it frightens me.' Death came for him in a moment. As he opened a bottle of wine, Stevenson felt something happen. It was a stroke, and within a few hours he was dead, his wish completed, dying in his boots. He wrote a requiem to be inscribed on his tombstone and I love its sentiments very much.

> Under the wide and starry sky,
> Dig the grave and let me lie.
> Glad did I live and gladly die,
> And laid me down with a will.
> This be the verse you grave for me:
> Here he lies where he longed to be;
> Home is the sailor, home from the sea,
> And the hunter home from the hill.

That is what I hope for, some years of decent health and good work, and an increasing sense of acceptance and perhaps peace. With all of the medical advances of the last century, we are living longer and longer, and I believe that we spend too much time denying death, frantically postponing it and consequently not thinking about what we'll do when it comes, not making preparations. If my time on the island had taught me anything, it was to come out of that sort of denial. In any case, age need not be seen as a continuing series of losses. Perhaps wisdom of a modest sort can be gained. For me, there is also the joy of my children, Adam, Helen and Beth, my grandchild, Grace, and, I hope, her cousins to come. That is not the eternal life so devoutly wished for by Cuthbert, but it is a version of it. When my son, Adam, was born, I passed on the delighted news to Andrew Cruickshank, a lovely actor and wonderful man who chaired the Edinburgh Festival Fringe when I ran it. He exclaimed down the phone, 'Immortal, my boy! Children make you immortal!'

12

Crossing the Causeway

I thought I might be bored. When my wife stays overnight at competitions with her lovely mare, I rattle around the farmhouse on my own, making a sandwich, half-finishing a glass of wine, watching the best bits of *The Godfather* for the hundredth time. I find myself in the evening flicking through TV channels or the pages of a novel, often going to bed early to shorten the day. But my time on Lindisfarne had passed very quickly. By doing so little, I seem to have done a great deal. Using my phone to take about five hundred photographs and filling a notebook, I felt myself become an observer. Rather than passing through, focusing on a goal or a destination, almost indiscriminately and certainly without a plan I began looking around the island, at its people and visitors, and thinking about rather than merely noting what I saw. I had taken a vow of silence. Most surprising was that after a time I did not miss conversation. In new places, I often talk to people I don't know and will probably never meet again. But here all my exchanges were brief and mostly practical. Perhaps closing my mouth opened my consciousness a little more and encouraged internal conversations, the sort of dialogue with myself I rarely have.

Pleasingly, my last day dawned bright and clear, and I had the morning to myself before my wife came to take me back across the causeway, back home and back to all that I love. Frustrated by the early closure of the priory on the first day, I had not gone back. In any event, I was much more intrigued by the much earlier monastery, by Aidan and Cuthbert and how they saw the island. In my notebook, I had scribbled 'Bog standard Benedictine monastery that hides what is really interesting'. Lindisfarne Priory looked to me much like many other ruined medieval churches. It was only its location that made it different from scores of others across Britain. But, as often on the island, my first impressions were wrong.

Made mellow and warmed by the morning sunshine, the sandstone of the priory church glowed. The view through the wrought-iron gate was almost painterly, and what is known as the Rainbow Arch, a surviving rib from the vaulting, completed a pleasing picture. I noticed that some of the pillars marching down the nave are decorated with chevrons and other motifs, like the massive pillars that hold up Durham Cathedral. However, more than architecture links the two churches.

When Aethelwine fled across the causeway to seek refuge from the armies of William the Conqueror, he re-made a broken link and unwittingly brought Lindisfarne back to life after the long centuries of abandonment. For that we should be grateful because if the priory had not been built, fewer people would come to Lindisfarne and the stories of Aidan, Cuthbert and the other saints would be less well understood. The brief return of the relics in 1069–70 were a prelude to much activity. The new Bishops-Palatine relied on the income and prestige derived from having Cuthbert's shrine at Durham, and when the old Anglian cathedral clergy were replaced by Benedictine monks after 1083 they decided to build a new church and set up a community on the island.

The original Congregation who came from Chester-le-Street could claim to be the legitimate heirs of Cuthbert, but the Benedictines could not. Instead they established their role as guardians of the saint and his cult by re-founding his monastery. They had also inherited the vast territories of the Haliwerfolc, the Holy Man's People, St Cuthbert's Land, and needed their ownership to be recognised and not disputed. Many of these possessions, such as Islandshire and Norhamshire, lay far from Durham but close to Lindisfarne.

By the early twelfth century, work on the new priory seems to have begun and it was completed in about 1150. A cenotaph, an empty tomb, was placed approximately where the original shrine of Cuthbert may have been. It stood close to the high altar at the east end of the church, the presbytery, behind a screen so that the laity in the nave could not see it – without paying something to the monks. This was set up partly to encourage pilgrimage and provide income but also to make clear the close links with Durham and its guardianship of the shrine of Cuthbert. The fact that the church was built first, and not, as was usual, the accommodation for the monks, underlines the importance the Bishops-Palatine placed on securing their legitimacy as heirs of Cuthbert's legacy.

The presbytery was in shadow and surprisingly cold out of the morning sun. An old Anglian cross-shaft had been placed close to where I imagined the cenotaph had been. I wondered how anyone could know the location of the original coffin after two centuries. And indeed, as the excavations on the Heugh have suggested, the saint's body may have rested in the chapel known as St Cuthbert's in the Sky. There is documentary evidence from the fourteenth century that the cenotaph had a painted statue of Cuthbert laid on top of it.

Facing south, the monastic buildings, the prior's lodging, the monks' dormitory, the storehouses and the stables caught all of the sun. There seems to have been no cloister but a large hall, and also the remains of a barbican gateway with a narrow, cobbled roadway running through it. This was a part of a series of defensive works made necessary by the outbreak of the long, intermittent war between Scotland and England that began in the late thirteenth century. On the south-eastern corner of the prior's lodging, facing the harbour, are the remains of a projecting tower. The Outer Court, whose walls are still mostly intact, would have looked like a farm stackyard, with animal pens, stables and haystacks. Because of the sunny morning, this was where most visitors walked, took photographs and sat down on benches.

I saw a man pulling behind him what looked like a small, wheeled suitcase with an extended handle. A tube ran from it up to his neck, around his cheeks and stopped under his nose with an outlet for each nostril, the sort of thing often seen attached to unconscious hospital patients. Over the ruined foundations, the cobbles and the uneven grass, this contraption was not easy to manage. When this man and his wife sat down gratefully on a bench that looked towards the Heugh, I broke my vow of silence.

'We come here as often as we can,' she said, and her husband explained that his lungs no longer supplied enough oxygen. The contraption did, by recycling and purifying the air, and it was what kept this man alive. He told me that he could not walk for long without sitting down and that the batteries lasted six to seven hours, enough time to get to the island, be there for two or three hours, and get home. Throughout our exchange, he smiled, and when asked why he made such an effort to come to the island: 'Peace,' he said. 'Nowhere else has this for me, and sometimes I even forget I have to trail this thing around.'

Near where the couple sat, stands a statue of Cuthbert very different from what was laid on the cenotaph in the old priory. Made by the same artist who carved the coffin bearers in St Mary's Church, it shows the saint seated and apparently at prayer, his hands clasped tight on his knee. It struck me as very evocative, especially the forming of the head. Like the man with the portable respirator, Cuthbert seemed to be at peace here, ready to meet his God.

As with the great abbeys of the Tweed Valley and the churches of the north of England, the centuries of war between Scotland and England were very destructive and saw decline at Lindisfarne Priory. From around ten monks in the twelfth century, numbers dropped to only two or three in the fifteenth. This handful of clerics probably held services for pilgrims and were attended by a group of servants. Despite the fact that lookouts were posted at the Snook, the landward end of the island's pan-handle, to warn of approaching soldiers, life was likely more than tolerable between periods of alarm. There are reports in the fifteenth century of two young monks behaving badly, playing dice, frequenting taverns and being 'swearers and utterers of prohibited jests'. By 1537 the priory had been closed by Henry VIII's commissioners and it began a long decline into ruin. Visitors still crossed the causeway, and written records and drawings they made showed that the church survived almost intact until around 1780, despite the lead having been stripped off the roof many years before. By the 1820s, the central tower above the crossing and the south aisle had collapsed and much of the stone was robbed out by local builders.

Walking back through the village, I could see where

some of the warm sandstone of the priory had been put to good use.

Before my wife arrived to collect me, I wanted to buy some of the excellent beans at Pilgrims Coffee House. The wind had begun to blow, but I sat down in a corner of the garden nevertheless. For no reason, and on the edge of a large group of chattering, animated people, I felt it again, but more powerfully: a sense of profound peace. Like exhaling a long breath, it seemed that all my cares slowly left me, like a receding tide. In that unlikely place, at that random moment, I knew I could look forward to the rest of my life without being afraid of death and look back at what had gone without regret. I have made many mistakes, hurt people I love, failed and committed many sins of omission, but I sat in the garden and thought differently about those dark times. I realised something absurdly simple: it may not have seemed so to the people around me, or indeed to myself, but I had done the best I could at the time. And what else could I have done? When death comes, that settlement with myself will help me face it. I knew that this was not a mood or a set of notions consciously assembled after a great deal of thought, but equally I understood that the moment would pass; this sense of peace would be fleeting. Something would happen to blow clouds over the sun and my attitudes would darken once more. But I knew at last that it was possible to come to an accommodation with the past and have real hope for a bright future. It was as if the island and the peace of Cuthbert had finally found their way inside my head, almost without me noticing.

I can only ponder more profound matters for a short time. I can't seem to consider life, death and everything for longer than about ten minutes. Perhaps that is because, unlike Cuthbert, Drythelm, Boisil and the other ascetics, I don't force myself to pray, stand up to my neck in freezing water or fast, using that effort of will to force out everything except the need to know the mind of God.

The starlings and predatory sparrows circled and hopped on the table, but I only had coffee so they moved on to other, better prospects with plates of scones or bacon rolls. Just as Cuthbert felt with the sea otters, the birds of Inner Farne and the eagle, this closeness was not an irritation to me, as it was four days ago. It felt good. Birds' behaviour, except when there are crumbs and titbits about, is difficult to read. These descendants of dinosaurs are enigmatic. From the bench at Emmanuel Head, I had watched eider ducks flying very fast straight out to sea. Where were they going and what was the urgency? Birds were everywhere on Lindisfarne and I wished I understood more about them.

Thinking about my time on the island, it occurred to me that the daily, weekly, monthly, annual, eternal battle to do good work, to take care of my family, to solve problems or at least mitigate their effect, does not seem to allow many breaks. That is where my stay on Lindisfarne had been different. By isolating myself, at least when the tide was shut, with no transport except my legs, I could do little about anything. I just had to let things go. Over the past days, I had become better and better at that – and while the whole journey from Old Melrose had not been an epiphany (I don't believe in instant conversions or blinding lights) so much as a good beginning, I could see it as the first few faltering steps in learning how to die. Now I knew it was possible to find Cuthbert's peace and I had discovered where it could be found. I decided I needed to keep coming back to the Island of Tides.

13

The Rock

Four weeks after I left Lindisfarne, I set out on the last part of the broken journey that had begun on Brotherstone Hill two months before. I wanted to go to Inner Farne, the tiny island that lies offshore Bamburgh Castle. Cuthbert died there on 20 March 687 and I thought it right to see where his story ended before this one did. Boat trips run from Seahouses until the end of October, and if I was to spend more than an hour or so on the island I needed to catch the earliest sailing at 10 a.m. That meant an even earlier start for me – up at 5.30 a.m. to see to the dogs and take Maidie, my little West Highland terrier, out for her morning walk in the dark.

A month after the moon had shone bright on Lindisfarne, it was full again and high in a clear sky. When I switched on the lights in the kitchen to make tea and feed the dogs, I heard an insistent tapping and scratching. It was the time of year when wood mice try to find a way into the warmth of the farmhouse, and I wondered about one causing chaos somewhere in the cupboards, but the dogs made no fuss. Then I saw it: a tiny goldcrest, its yellow-streaked head visible in the light by the window. Hanging on to the frame, it was tapping at the wood to flush out the tiny insects that had taken refuge there.

The moon shadows behind the hedges and woods were black-dark, but the tracks were bathed in a monochrome light. Maidie and I had no trouble finding our way. The open sky and a stiff north wind made the early morning bitterly cold. Once out of the lee of the trees that line our track and onto the old road that leads down to the valley floor, we both shivered and walked more briskly. The light came first from the west, but the high, pale moon against a grey sky was slowly matched by a gathering of pink in the east, as the sun began to climb. I thought of rushing around the bay to watch it rise at Lindisfarne Castle.

From the old road, there are long views down our little valley and I saw dark morning lights: a searchlight sweeping over the high ground, perhaps someone out early, lamping for foxes. There were headlights moving up the Thief Road. About three miles away, and once a track used by cattle thieves, it leads up into the hills to the south. A few moments later the headlights swung across the moorland plateau and disappeared.

By the time I hung up my jacket in the porch and went to make some breakfast, the colours of the land were slowly coming alive, the sun's first rays lighting up the tops of the trees around the farmhouse.

* * *

After turning off the A1 south of Berwick, the road quickly finds the coast and climbs up past a links golf course before a stunning reveal takes place. Around a corner, the breath-catching, mighty mass of Bamburgh Castle suddenly comes into view. Towering over the little village at its foot and looking out over the sea beyond, it is epic in its scale and setting, surely a fortress that reflected the power of the kings who built it. Enhanced by the restoration and remodelling commissioned by William Armstrong more than a century ago, it must be the most dramatic castle in Britain, even more impressive than Edinburgh. History seems to seep out of its stones.

Three miles further south lies an eye-watering contrast with all that grandeur. From a former life as a fishing port and safe harbour, Seahouses has been rebuilt as a miniature Blackpool. On three sides of its central roundabout are chip shops with seating for hundreds and long takeaway counters, an amusement arcade whose dark interior winks with lights, a bargain clothing shop, a crazy golf course round the corner, more chip shops and, masquerading as a gift shop, an Aladdin's Cave that seems to defy every rule of the retail trade. I love it all.

Billy Shiel's Boat Trips sail from the end of the harbour pier and, despite the sunny, windless day, there seemed to be only three passengers boarding the *St Cuthbert II*. Older than me and with shamingly good English, a German couple sat in the stern as we pushed off from the moorings. Most people make these trips to see the bird life, especially in the breeding season, and the seals, but the puffins and the other migrating species were long gone. My plan was to be put ashore on Inner Farne and be picked up by a later sailing. I wanted to spend as much time on Cuthbert's island as I could. But first we were to make our way out to the farthest islands in the little archipelago. Depending on the tides, there are either fifteen or twenty, some of them little more than rocks peeping above the waves. As we nosed slowly out of Seahouses harbour, past the high concrete sides of the pier, I felt a familiar frisson of fear. The pier rises sheer and monumental out of the harbour like a bulwark against the deeps of the sea. I swim like a brick and if I fell overboard

I would surely drown. I noted where the life belts were stowed.

Mercifully, it was flat calm, and when we cleared the harbour entrance to plough out into the open sea the low rocks of the Farnes came into view. They look deceptive. As we sailed closer, the south-western cliffs of the tiny island of Inner Farne rose up out of the water, high and jagged. Parts were vertical stacks, almost detached from the cliff face. The skipper took us past Cuthbert's island to sail through a very narrow channel between Little Scarcar and Big Scarcar, two rocks that must have been singular undersea crags with very steep sides invisible under the water. If the boat had had wing mirrors, they would have scraped against the low cliffs. Several cormorants perched above us and the German couple took photographs. The birds reminded me of Graculus, the talking messenger-bird from *Noggin the Nog*, a TV cartoon I loved when I was a child. The German man rarely removed his arms from around his wife's shoulders or her waist, and while it looked like genuine affection I wondered if she was as nervous a sailor as me.

In the distance I could see Brownsman Island, the site of the original, relatively primitive Longstone Lighthouse. Beyond it is the modern version, on top of its characteristic red-and-white tower, the light I saw from Lindisfarne. The lighthouse features in a very famous and dramatic story. At about 4 a.m. on 7 September 1838 tremendous seas and gale force winds drove the steamship *Forfarshire* onto the deadly rocks of the Farnes. The ship foundered on Big Harcar, between Brownsman and Longstone, and immediately broke in two. Forty-three of those on board were plunged into the deeps to their deaths. As dawn broke through the storm clouds, Grace Darling, the daughter of the keeper of the Longstone Lighthouse, saw the outline of the wreck and searched through the swirling mist with a telescope for any signs of survivors. At about 7 a.m. she saw some movement on the black, sea-lashed rocks and in an open boat with her father she rowed out into the mountainous seas to attempt a rescue. Taking a leeward and lengthy course, they made two trips and nine people were saved.

A week later the *Newcastle Journal* carried a full account of the episode, which stressed that it was Grace who had persuaded her father, William Darling, to put to sea in the storm. The story was picked up by the *London Times* and its report asked, 'Is there in the whole field of history, or of fiction even, one instance of female heroism to compare for one moment with this?' Grace Darling immediately became the object of hysterical admiration. Tourists came to the Farnes to see her, to beg for a lock of her hair. Offers of marriage poured in, awards, decorations, a London theatre offered her a starring role in *Wreck at Sea* for £50 a week. William Wordsworth was moved to write a eulogy and she was made 'National Heroine of Japan'. There is now a Grace Darling Museum in Bamburgh.

The museum is also a monument to a story that spun out of control, a very early example of how newspapers can create a perception that obscures much of the truth. What really happened as the sea boiled around the deadly rocks was buried under all the adulation. As with many jobs in the nineteenth century, lighthouse keeping involved the whole family, although only one man was paid. When the *Forfarshire* struck Big Harcar at 4 a.m., Mrs Darling was on watch while her husband slept; the day before, Grace had helped her father to lash their rowing boat safely before the coming storm. Her brother, William Darling Jr, was also part of the family business. And so it was by no means unusual for Grace to be scanning the seascape with a telescope. In *Grace Darling*

– *Her True Story*, written by her sister, Thomasin, and a Mr Daniel Atkinson, it is made clear that William Darling Sr made the decision to attempt a rescue because he believed that neither the Seahouses or Bamburgh lifeboats would launch in such a fierce storm.

Darling's rowing boat was the only chance for the survivors of the *Forfarshire* and, since his son was away, only Grace was available to row with him. What she did was indeed heroic, but not quite as heroic as the newspapers made out. In fact, Grace seems to have been embarrassed at all the fuss and it may have contributed to her early death in 1841, three years after the rescue. The seabed around the Farnes is littered with the hulks of hundreds of wrecks and the bones of thousands of sailors who drowned, thrashing and struggling in the icy waters. These were not thoughts to comfort a nervous sailor.

On a day of bright sunshine and flat calm, the Farne Islands looked very different from the night of the fatal storm that drove the *Forfarshire* against the rocks almost two hundred years ago. The skipper slowed the engine and took the boat alongside a series of low ledges close to where the ship had broken in half, so that we could see the basking seals close up. Unlike the howling choir on the sandbanks off Lindisfarne, these were much bigger North Atlantic seals, rather than the common or grey seals. Some, said the skipper in his excellent commentary, weigh three-quarters of a ton. Warming themselves on the rocks, they looked it: huge and ungainly, occasionally shuffling around with great effort. Two pups stayed close to their mothers because gulls often attack them, going first for their eyes. The big seals turned their soulful, almost mournful faces to look at us looking at them.

We sailed between Brownsman Island and South Wamses,

a strange name, to the north-western side of the archipelago, and then turned for the landing stage at Inner Farne. To my surprise, there were no cliffs and the shore rose up very gradually out of the sea. Beside the concrete jetty, there was a small beach. I agreed a pick-up time with the skipper and, like a frail, old person, I was helped ashore by many hands. The island is in the care of the National Trust for England and I was greeted by very friendly young rangers. All of them had been delighted to come and work and live on Inner Farne for most of the year because of a keen interest in wildlife. One young man had a pair of binoculars around his neck and, as he talked to me, he often looked through them at something distant. None of them seemed to know much about Cuthbert, and when I asked to see the remains of his hermitage, no one knew where to look.

Built over the site of where it might have been is a fifteenth-century tower, its function defensive. Much like Lindisfarne Castle, it protected strategic sea-roads during the long war between England and Scotland. The tower forms the north-western side of a small courtyard that also includes a chapel that was dedicated to St Mary and whose ruins were adapted to create a small visitor centre. It was used by women only. Opposite is another chapel that was rebuilt in the fourteenth century and dedicated to St Cuthbert. Inside I found elaborately carved wooden pews that would not have looked out of place in a cathedral and a tripartite stained-glass window with a portrait of Cuthbert holding the head of St Oswald, flanked by St Aidan and St Aethelwold, a later hermit on Inner Farne. There is also a monument to Grace Darling, with some overwrought lines below her name that begin: 'Oh! That winds and waves could speak of things that their united power called forth from the pure depths of her humanity.'

On the floor were some carved stones that were medieval in style, but in and around the courtyard (in the places where I was allowed to venture) I could find no sign of Cuthbert.

Both the Anonymous *Life* and Bede include a good deal of detail about what the saint built on Inner Farne, probably with help from the brethren on Lindisfarne. Aidan had retreated to the island for short periods when he was bishop and there may already have been a shelter of some kind. Near a well (one of the rangers told me the tower had been built over a well) and on the more sheltered north-eastern shore by the beach was clearly the place to build. Cuthbert appears to have dug down into the soil of the island to form what archaeologists call a '*grubenhaus*', a building method imported by Anglian and Saxon settlers. To form walls, he piled up uncut stones and God sent driftwood for the rafters of a thatched roof. The tale of the naughty nesting ravens confirms that method of making his cell weathertight.

Next to his cell, and a well which was discovered in an Old Testament flourish that recalled Moses striking the rock, was his oratory. Bede added more detail about what was a large compound:

> It is a structure almost round in plan, measuring about four or five poles [about eighty feet] from wall to wall; the wall itself on the outside is higher than a man standing upright; but inside he made it much higher by cutting away the living rock, so that the pious inhabitant could see nothing except the sky from his dwelling, thus restraining both the lust of the eyes and of the thoughts and lifting the whole bent of his mind to higher things.

Sanitation was a surprising consideration for a hermit alone on a sixteen-acre island, but I imagine Cuthbert was fastidious in his habits. Here is a fascinating passage, again from Bede, with some wonderfully restrained language about the practicalities of the solitary life:

> The very sea, I say, was ready to do service to the servant of Christ when he needed it. For he was intending to build a hut in his monastery, very small but suited to his daily needs; it was to be on the seaward side where the hollowing out of the rock by the washings of continual tides had made a very deep and wide gap; flooring had to be placed under the hut, and this had to be twelve feet long so as to fit the width of the gap. So he asked the brethren, who had come to visit him, that when they were returning, they would bring with them some timber twelve feet long, to make a flooring for this little house.

But the brethren forgot and instead the required timber for the toilet was supplied by driftwood that happened to be exactly twelve feet in length and fitted the place where the tides would flush the loo.

Cuthbert's cell and oratory have almost certainly been built over by the later chapel and tower, but the foundations of his guesthouse for visitors may have been used to build a fisherman's hut by the jetty. It stands at some distance from the round compound. Instead of poking around the courtyard, I decided to walk around the little island. Perhaps a sense of Cuthbert's presence might be found elsewhere.

St Bartholomew of Farne lived on Inner Farne for forty-two years, leading an extremely ascetic life in the second half of the twelfth century. He wore skins, slept on rocks, lived on bread he made from corn he grew on the island and milk from a cow, as well as fish he caught. Bartholomew's habit was to stride around the island singing psalms in a ringing voice. In another sense, he led a sheltered life. When a woman strayed from St Mary's Chapel into St Cuthbert's, the horrified hermit is said to have fainted. Famous and much admired, he was visited by wealthy people who probably made gifts to the church and sought his blessing. When Bartholomew died in 1193, he was buried on the island, but I could find no sign of his tomb. Or anything else that took Inner Farne's story further back than the later Middle Ages.

To protect the nesting birds, visitors are bound to follow the roped-off boardwalk paths around Inner Farne. What I had at first thought to be rabbit holes were of course puffin burrows and in the surprisingly deep soil there were hundreds, maybe thousands. I walked up to the highest point of the island, where a lighthouse had been built in the nineteenth century, and was struck by how close it is to Bamburgh Castle. The Northumbrian kings who looked out to sea could not miss Inner Farne, the place where, according to the Anonymous *Life*, Cuthbert did battle with evil, and 'a place where, before this, almost no one could remain alone for any length of time on account of the various illusions caused by devils'. The biblical reference would have been unmistakable. In the Gospel of Luke, a beggar covered in sores called Lazarus lay at the gate of a 'rich man dressed in purple and fine linen' and lived off the crumbs from his table.

Beyond the lighthouse was a sheer drop to the sea, and the stacks I had seen from the boat were very dramatic. I wondered if there were prayer holes in the cliff, like those on the Heugh on Lindisfarne. Amongst the shelved rocks where I stood was a beautifully formed natural basin where rainwater had gathered, similar to what I had seen on top of Cuddy's Cave. Swimming offshore were two eider ducks, known locally as Cuddy's Ducks. When he lived on Inner Farne, the saint insisted that the birds not be attacked or their eggs taken. This very early conservation initiative was something the young rangers did know about the hermit.

Neither the Anonymous *Life* nor Bede's note the date when Cuthbert laid down the cares of office on Lindisfarne and came first to live alone on the island, but his last period of only a few months is well documented. After Christmas 686, and despite what seems to have been a worsening, debilitating illness, Cuthbert sailed the winter seas to Inner Farne. His trust in God must have been absolute. In the place where he had prayed, sung psalms and fought devils, he wanted to meet his Maker. From Bede's account, it appears that he was often not alone. In the last week of his life, Herefrith, who became abbot of Lindisfarne, came with other brethren and much of what Bede wrote is his verbatim recollection. As Cuthbert's condition deteriorated and 'the stress of his sickness took from him the power of speech', he retreated to his oratory. Herefrith remembered:

I entered in to him about the ninth hour of the day and I found him lying in a corner of his oratory, opposite to the altar; so I sat down by him. He did not say much because the weight of his affliction had lessened the power of speech.

After Cuthbert had 'passed a quiet day in the expectation of future bliss', his condition worsened once more. On the night of 20 March 687, he died:

Without delay, one of them [the brethren] ran out and lit two torches: and holding one in each hand he went on to some higher ground to show the brethren who were in the Lindisfarne monastery that his holy soul had gone with the Lord: for this was the sign they had agreed amongst themselves to notify his most holy death.

When the watcher on the Heugh, five miles to the north, saw the pinpricks of flaring torchlight on Inner Farne, he may have lit an answering beacon. Cuthbert had told Herefrith that he wished to be buried on Inner Farne, 'where, to some small extent, I have fought my fight for the Lord, where I desire to finish my course, and where I hope I shall be raised up to receive the crown of righteousness from the righteous Judge'. He was also anxious about pilgrimage, showing a keen awareness of how cults develop and about 'the influx of fugitives and guilty men of every sort' to Lindisfarne. Eventually Herefrith managed to persuade him to allow his body to be taken back to the monastery. The earlier Anonymous *Life* makes no mention of Cuthbert's reluctance, whereas Bede spends a page rehearsing the discussion and in it there is a powerful sense of reading history backwards. By the time he wrote his account, the cult of Cuthbert had flowered, the *Lindisfarne Gospels* were complete in all their glory, and no doubt many pilgrims were making their way to his tomb and contributing gifts to the monastery and the church in general. It probably appeared seemly to Bede for Cuthbert to have shown reluctance to accept in death the fame he feared in life.

Here is the succinct passage from the Anonymous *Life* on the day it became clear that Cuthbert had been elevated to sainthood:

After eleven years, through the prompting and instruction of the Holy Spirit, after a council had been held by the elders and licence had been given by the holy Bishop Eadberht, the most faithful men of the whole congregation decided to raise the relics of the bones of the holy Bishop Cuthbert from his sepulchre. And, on first opening the sepulchre, they found a thing marvellous to relate, namely that the whole body was as undecayed as when they had buried it eleven years before. The skin had not decayed nor grown old, nor the sinews become dry, making the body tautly stretched and stiff; but the limbs lay at rest with all the appearance of life and were still moveable at the joints. For his neck and knees were those of a living man; and when they lifted him from the tomb, they could bend him as they wished. None of his vestments and footwear which touched the flesh of his body was worn away. They unwound the headcloth in which his head was wrapped and found that it kept all the beauty of its first whiteness; and the new shoes, with which he was shod, are preserved in our church over against the relics, for a testimony, up to this present day.

To my eye, Inner Farne is not beautiful or indeed atmospheric, like Lindisfarne. My time on the island was brief and the day sunny and calm, but no spirits seemed to inhabit it, nor any devils. Even though I was the sole visitor and was not distracted, I found the rock to be a historical blank and gained no insight from walking where Cuthbert had walked. There was nothing there except the documented bleakness of the hermit's weary death and the management of his journey to sainthood. In both literal and metaphorical terms, even the account of his disinterment is a story of manipulation. The image of the Lindisfarne monks bending his lifeless body, like a floppy doll, is not attractive, even though it must be apocryphal.

I waited on the edge of the little beach by the jetty for the boat to collect me and thought about how Cuthbert had met his end. In Bede's account, it reads like a recital of miseries, many of them self-inflicted. When he resigned his bishopric, or was pushed out, Cuthbert seems to have known he was very ill and, in an era without much medical succour, was probably suffering a good deal of pain. His faith and trust in God took him to Inner Farne because that was where he had fought the good fight hardest and where it would be good to die. But for a sick man lying on a rock scoured by the icy winter winds in the long nights of January and February, it was a time of extreme privation, the worst pain he had ever endured. At first I found it very difficult to discern any lessons or examples in all of that misery.

When at last the boat arrived and I clambered aboard with more helping hands, the trip back to Seahouses seemed to take forever. As I looked over my shoulder at Inner Farne, I realised that Cuthbert returned there to take up the hermetic life because he wanted to face death alone. And he was right to do that. We all face death alone; no one goes with us into the darkness. But it was the suffering that upset me, the painful conclusion of an exemplary life. On my journey from Brotherstone Hill and Old Melrose, and through my attempts to understand why he did what he did, I had felt that sometimes I walked beside Cuthbert, and I found the story of his death vexed me. What I seek is not something philosophically complex. I am no philosopher. I want to be at peace with myself when I come to the moment of death and I hope that illness will not make it painful. For all his prayer and all his evident goodness, Cuthbert seemed to me to meet death with little peace. Often unable to speak, having collapsed, enduring the bitterness of winter and all the time negotiating with Herefrith about where he might be buried, Cuthbert's death appeared to be more of a release than a triumphant ascent to heaven in the arms of angels.

Epilogue

Godless

The first frosts of winter had come. Late home from Edinburgh the night before, I looked up at an open sky. Orion with his belt, sword and square shoulders stood guard in the east, the Plough was upended above the old sycamores in the Top Wood and the Milky Way flowed its neck-craning course across the black of the heavens. Quiet, fast asleep, the farmhouse, stableyard and the home paddock were bright with starlight. When I was out with the dogs in the early morning, I felt a sharp, headachey cold in the air and in the half-light before dawn the fields were white. Later I looked at the gauge on the woodstore: minus seven.

By the time Maidie and I had walked up the track, there was enough light to make the stars fade to blue, even though the sun had yet to breast Greenhill Heights. Over in the west, I saw the sheep waiting patiently on Howden Hill to warm themselves in the first rays. The rays were already shining on the hills above the Ettrick Valley, making them glow pale purple in the windless morning. The long, hot summer had laden the apple trees and the cherries with abundance. On one small tree I lost count of the number of apples after 200. Most will drop unused as windfall, a waste. The frost had made the cherries glacé, but the birds will savour their goodness nonetheless. Even if the winter begins harshly, there will be plenty of berries, cherries and apples for them, but in January the robins will become friendly, hopping close in the stableyard, daring to dart in and out of the feed-room in search of scraps.

When Maidie and I reached the Windygates, the highest point of the track, cracking the crusts of ice in the puddles, the sun came quickly, just as it had over the grey horizon of the North Sea beyond Lindisfarne. It was magical. The birches still had most of their tiny, filigree leaves, the sycamores were not yet bare, and the larches were pale yellow, waiting for a stiff breeze to take their needles. When the sun came, the trees glowed golden, the white film of frost around them evaporated and the land came alive. Many years ago, I visited an old priest on South Uist in the late autumn and Father Angus told me how he felt the seasons shift. When he was a boy growing up in the islands, he put away his boots in the spring and walked and ran barefoot. 'I could feel the land waking under my feet, the soil warming and the light returning. And at the end of autumn, I could feel it beginning to die again.'

Between returning from Inner Farne and writing this, all of my resolutions and musings on how to face death suddenly came into sharp focus with some blunt possibilities. After several weeks of developing symptoms that I thought were nothing much, maybe a passing infection, I went to see a doctor, anxious about prostate cancer. It is becoming very common, with nearly 50,000 diagnoses in Britain each year and growing. I gave the doctor a sample, he tested it immediately, and, saying he had found some blood, he set up more tests and an appointment with a specialist at the local hospital. I suspected his speed and efficiency were precautionary rather than something to be alarmed about. My blood test would tell. If I had prostate cancer, the degree of it, what is known as a tumour marker, would show up.

I had to wait six days for the result, partly because the laboratory was closed over the weekend.

During that anxious time, I forced myself not to ignore, not dismiss this as an anxiety about something that had not happened and let it lurk at the back of my mind. And equally, not to be melodramatic. I rehearsed as calmly as I could what I would say if the result came back positive, confirming that I had cancer. Most important would be the conversations with Lindsay and my children. By that time, I hoped I would know something about how advanced any cancer was by the marker rating and I spent time doing what friends advise you not to do, researching on the internet. I thought it best to confront this possibility squarely and the best way to do that was with good information.

Over those days of waiting, I thought a great deal about my time on Lindisfarne and my journey in Cuthbert's shadow. It occurred to me that on Inner Farne he was tired of dying, impatient to be away, to join his God. All of the weary negotiations with Herefrith ended when, it seemed to me, Cuthbert gave up and gave in. It was a sad process I had seen for myself. After my father died on that snowy evening in February 1986, my mum gradually began to withdraw. Even though she complained, particularly to my sister Barbara, about looking after my dad when he was disabled by strokes, she found life without him difficult. Although she loved her grandchildren and her children, I think she felt her life had little purpose, and when she died it was because she had had enough. I wondered if I would grow tired of dying, especially if the cancer was advanced. And, of course, these dark thoughts were intermittently blown away by sunlit gusts of optimism. The doctor's haste was nothing more than very welcome efficiency, the NHS working well. Of course it was.

While it took place after days of shivering misery on Inner Farne, Cuthbert's departure and his negotiations with Herefrith were not all negative. He was doing something sensible: thinking about his death and its consequences for others, planning, considering his legacy. It is something we should all do, but preferably not in the shadow of imminent death. If my blood test came back with a high marker, meaning the cancer had spread, then I too would have to make some decisions, but I decided to proceed slowly, one step at a time. Sometimes I find life wearying, but at sixty-eight I was not ready to go, if the news from the laboratory turned out to be bad. It might not be, but I could not ignore other possibilities.

On the journey to Lindisfarne, there were moments when I recognised that Cuthbert had given me gifts and I wanted to think about them. Up on Brotherstone Hill, after I had managed to find it, I remember being overwhelmed by the glories of the Border landscape, my native place. The hills and river valleys were not just stunning scenery but places enriched by experience, made even more beautiful by events and not only things that happened to me. History rumbled down these roads, and lives were passed tilling those fields and tending the flocks on the hills. This place is not only a series of geological accidents but also somewhere made by the back-breaking labour of farmers and farm workers over millennia. Those heaps of years should not be remote, a time of long processes and small events that happened to other people. My family has been in the Tweed Valley for many centuries, their lives unremarked by historians, and they and countless others made the land look the way it does. Long before I walked its roads and tracks, they walked them too. For me, history is not something found in books or in TV programmes, it is deeply personal. Up on Brotherstone Hill, I understood that.

As I walked the banks of the Tweed and Till and across the moors and the Kyloe Hills, I looked at the changing landscape, but also noted how little its essence had altered since Cuthbert's time. History is about change and loss, but also about continuity. There were times I saw sights Cuthbert had seen and often I felt I could reach out across the centuries to take his hand. He has achieved immortality of a sort, living on through the pages of history, his exemplary life inspiring many pilgrims to keep his memory bright. Mine will fade much more quickly, probably dying with the generation after my grandchildren's. But it is enough to know that my children and theirs will remember me and understand something of the continuity I cherish as the generations unfurl after them.

Cuthbert's love of animals and his longing for the coming of the Second Creation also had lessons. To heal itself and avoid the coming cataclysm of climate change, our broken and divided world is going to have to change. We will need to return to having respect for the rest of Creation and not see it only as a background, a context for human dominance. As President Macron of France told the US Congress in 2018, there is no Planet B, and a Second Creation of a secular sort must be undertaken. Despite his speech and sentiments, political leaders will not take initiatives until it is too late. The atavistic chaos in the United States and elsewhere means that communities and individuals have to begin the necessary process of change from exploitation to respect for our world. Cuthbert loved it because it was God's Creation and we should love it because there is no alternative. The central difficulty is the disunity of purpose and policy across what seem like impossible and impassable divides. While building vast cities that are the toxic engines of pollution, the Chinese appear to be bent on growth at all costs, on dominance even, while other Asian economies are expanding at dangerous and poisonous rates. The President of Brazil seems poised to sanction attacks on the Amazonian rainforest. All of these nations want the standard of living the West has long enjoyed and are unlikely to listen to lectures on climate change and its causes. This is nothing less than the greatest crisis in the world's history and all we as individuals can do is live simpler, less-consuming lives.

Throughout his adult life, Cuthbert craved solitude so that he had no distractions in his endless quest to move closer to God. He went to extraordinary lengths to achieve a hermetic life. By comparison to the cold and hungry years he spent alone, my time on Lindisfarne was piddling, but it did teach me something important. I had never in all my life spent such a long time alone and largely silent. To my surprise, I not only became comfortable in my own company, but I also enjoyed the peace that descended when I sat alone on the Heugh or out at Emmanuel Head. A simple gift from Cuthbert, but one I shall value in the life to come.

On the journey to the Island of Tides, I thought a good deal about the dead, about Hannah and the life she did not have, and my mum, dad and grannie, and the life they gave me. Instead of shedding more tears (all too easy for me), I found myself learning another lesson. So that we can find comfort in our sense of our lives, we should keep the dead close. They live on in us just as we shall live on in the lives of our children and grandchildren. It is a continuum, a version of an afterlife I cherish.

Part of my mission on the road to Lindisfarne was to focus on the good things in life and try to pull away from darker matters, mistakes, self-criticism, regrets and pessimism. I had also resolved to forgive those people who have hurt me over the years and I will not resile from that.

Reading back through this manuscript, I was struck by how often I wrote about death. Knowing it is almost a taboo subject, I was tempted to edit out some of these thoughts. I resisted for a simple reason: it was an important part of my purpose in walking to Lindisfarne, to confront death and come to an accommodation with myself that would help me when it comes. I am not sure I have yet succeeded, but at least the process has begun. And thinking about my death has certainly had an effect on how I want to live the life left to me.

The night before I wrote this, I interviewed the former Scotland rugby player Doddie Weir in front of an audience of more than five hundred about his recently published autobiography. At Christmas 2016, he was diagnosed with motor neurone disease. He was only forty-six at the time. In what sounded like a cold and clinical manner, the consultant told Doddie that he would not walk into the hospital in a year's time. After the shock had subsided, all of his competitive, rugby-playing instincts shoved their way to the front of his mind and, appalled at the lack of available therapies and research, Doddie launched a charitable foundation to raise money. At the time of writing, he has given more than £1 million to help those who suffer from this terminal condition and also to pay for further research. And almost two years after diagnosis he is still walking.

In front of a packed audience of fellow Borderers, Doddie put on a bravura performance, even though he was clearly exhausted. I had seen him five months before and it seemed to me that his condition was deteriorating. He was confronting death with humour, some of it dark, and with action, all of it positive. Doddie knows he will die soon, but he is determined to go down fighting. In between laughing at his jokes, the audience repeatedly cheered him, applauding his immense courage. Perhaps it is the certainty that his massive six-foot, six-inch frame will fail and fall that makes his tragedy more poignant. There were moments, little more than glances, when I saw the hurt and the sadness behind his smile. Doddie's strategy for confronting death works, but not all the time. And for me that too was a lesson. Even if I reach a settlement with myself, there will be days, perhaps weeks, when it falters.

On the same day I met Doddie, I had news of my blood sample. It was mostly good. I do not have prostate cancer and that is a mighty relief, but there is something wrong. I will need to have more tests to investigate whatever it is that is giving me frequent pain in my lower abdomen. I suspect we will never find out, and it will clear up anyway. When I shook Doddie's hand at the end of the evening, I almost felt guilty that I had had a reprieve and that there would be none for him.

Up on Brotherstone Hill, late on an autumn morning, I looked east down the valley of the Tweed under skies that threatened rain. A southerly breeze was keeping the weather moving but overnight rain had made the grass too damp for sitting. Standing in the lee of the larger of the Brothers' Stones, I felt I knew more than I did when I last climbed the hill. On my journey I had learned a great deal about Cuthbert and there seemed to be much that he could teach me. I know that his peace will come to me gradually, and in times when it does not I now know where to find it.

Like the people whose names are commemorated on the benches, I have come to love Lindisfarne. As they did, I see it as a haven, a place to run to when life overwhelms me, a place to grieve and a place to find a simple joy in the glories of the sea, the sky and the land. When I walked back down off Brotherstone Hill, I realised that this time I had found the right path.

Acknowledgements

I want to begin by thanking Simon Thorogood, my editor, for his faith in this unlikely book, and for his astute suggestions on how to improve it. Debs Warner has done a lovely job in editing the text and removing some of the grit and smoothing some rocky passages. Thank you both.

For twenty years, my agent, David Godwin, has been much more than that. Every miserable scribbling drudge needs encouragement, tissues to dry their tears and some firm guidance. No one does all these things better than David.

Walter Elliot read every word before the story went anywhere else. His generosity in sharing his encyclopedic store of knowledge and lore, his gentleness in pointing out blunders and infelicities, and his passion for the stories I wanted to tell – all of these kindnesses were central to my enjoyment in writing about this journey and about Cuthbert.

To the Island of Tides is for Richard Buccleuch in inadequate acknowledgement not only for his unwavering support for what I do but also for all he does to make the Borders a better place. Our talks at Bowhill over many years have sustained me, made me laugh and made me a believer in the fundamental decency of the human spirit. Richard is

To the Island of Tides

not grand, but instead a great man, and this journey and
this book are dedicated to him with love and friendship.

Alistair Moffat
Selkirk, March 2019

Index

Abrahamsen, Tormod 202–3
Ad Gefrin 161–2
AD time system 100–1
Aebbe, Abbess 77
Aelfflaed 241, 242–3, 244
Aelfwald, King 263
Aelle 218
Aethelfrith 100, 101, 218–19
Aethelthryth (Etheldreda),
   Queen 153, 245
Aethelwine, Bishop 271, 277
afterlife 68–9, 72–3
Aidan, St , xvi, 12, 198–9,
   200, 266
   and Inner Farne 290
Alcuin of York 262–3, 267
Aldfrith, King 151, 152, 242,
   244, 245
Aldhun, Bishop 269
Aldred 250–1, 268
Alexander I, King 60
Alexander II, Pope 152
Alexander III, King 132
America 91
Angles 117–18, 171–3, 175–6,
   179

animals 77–84, 248–9; see
   also birdlife
Anthony of Egypt, St 45–6,
   191, 238
Armstrong, Andrew 16
Armstrong, William 217,
   218, 285
Arthur, King 117, 118, 178
ascetics 45–6, 49, 95
Athanasius of Alexandria,
   St 45–6
Audrey, St 153
Avenel, Robert, Lord of
   Eskdale 65, 68

Bamburgh xv, 99, 173
   and Castle 10, 217–18,
      284–5
Bannockburn 135
Baptised, the 99, 101, 179
bards 177–9
Bartholomew of Farne, St
   291–2
Basil, St 64
Battle of the Standard 135
Beal 3–4

Bede, St xv, 16, 29, 152
  and Drythelm 85–6
  and Eata 214–16
  and *Ecclesiastical History
    of the English People*
    172
  and Inner Farne 290–1,
    294
  and *Metrical Life* 94
  and Old Melrose 33–4, 37,
    62–3, 66, 90
  and *On the Reckoning of
    Time* 100–1
Beltane 27, 48
Bemersyde 37–8, 40–2
benches 201–4
Benedictines 277–8
Benrig Cemetery 107–8
Bernicia 99, 218–19
Berwickshire Naturalists'
  Club (BNC) 15–16, 35
Billfrith 251
birdlife 110–11, 140–1, 282,
  293
Birgham 131–3
Black Death 87
Blodmonath (Blood Month)
  101
bog bodies 49
Boisil, Prior 12, 63–4, 86–8,
  89, 95
  and Chapel 107, 108
Boyd, Margaret 272–3

Braudel, Fernand x
Brendan the Navigator, St
  91–2
Bridge Well 234, 245–6
Brisbane, Sir Thomas 113
Brothers' Stones 18, 21–3,
  28–9
Brotherstone Hill 16–20,
  21–5, 28–9
Brownmoor 14–15
Brownsman Island 286, 288

Cadrod, Lord of
  Calchvynydd 99
calfskin 253–4, 255
Candida Casa 102, 180
Canon Tables 252
Carham 130–1, 133
Carlisle 101–2
Carnais 223–4
Catterick 99
causeway 4–5, 6–7, 181–2,
  185–7
Cellach, Abbot 263
Celtic Church 94, 95, 151,
  219–20
Celts 26–8
Chester-le-Street 250, 268
Cheviot Hills 23
Christianity 180–1, 193–4,
  255–6
Church of Rome 94, 95,
  151, 219–20

Cistercians 61
Clearances 223–4
Cliftonhill Farm 126–8
climate change 303–4
Cnut, King 269
Coldinghamshire 155
Coldstream 136
Colgrave, Bertram 16
Colman, Bishop xv
Columba, St xv, 91–2
Columbus, Christopher 91
Congregation of St
    Cuthbert 266–9
conversion 101–2
Coquet Island 241–3
coracles 92–3
crosses 42–3, 232–3
Crossman, Sir William
    216
Cruickshank, Andrew 275
Crystal Well 105, 107
Cuddy's Cave 160–1
Cumbria 101–2, 174, 176, 180
curraghs 92–3
Cuthbert, St xi–xii, xiii, xiv,
    xv, xvi, 11–13, 30
    and afterlife 68–9
    and animals 77–80
    and banner 135, 141
    and Benedictines 277–8
    and Brothers' Stones 28,
        29
    and Carlisle 101–2

and Church of Rome
    219–20
and churches 198–9
and coffin 195, 196
and Cuddy's Cave 160–1,
    162–3
and cult 60–1
and death 293–4, 296–7,
    301–2
and dunes 237–8
and Eata 153–4, 214–16
and Ecgfrith 241–5
and flight 93–4, 98, 99,
    114–15, 121–3
and Inner Farne 240–1,
    289–91, 292
and lives 16–18
and Monksford 52, 53, 55,
    56
and Old Melrose 32–7, 47,
    62–3, 86–8, 89, 90
and relics 266–7, 268–9,
    270, 271
and sainthood 294–5
and Synod of Whitby
    95–6
and Vikings 144
and Wilfred 151–2

Darling, Grace 9, 286–8, 289
David I, King 60–1, 122, 135
death 69–75, 87–90, 272–3,
    305–6; *see also* afterlife

Deira 218–19
Demarco, Ricky 246–7
Dere Street 153–4
Desert Fathers 45–6
Devil, the 122–3
Dionysius Exiguus 100–1
Doddington Moor 162–3
Domhnall Ban Crosd 223–5
Donan, St 263
'Dream of the Rood, The'
    192
Druids 48, 95
Dryburgh Abbey 35, 41,
    42–3, 44–5, 47–8, 50–1
Drythelm 85–6
Dublin 264
Duddingston Loch 105–6
Dun Airchille 23
Dunnichen 245
Durham Cathedral xi, xiii,
    xvi, 61, 269–70, 256

Eadfrith, Bishop xv, xvi
    and *Lindisfarne Gospels*
        250, 251, 254–5, 256
Eadulf of Bamburgh, Earl
    131
Eardulf, Bishop 268
Earlston Black Hill 23
Easter 94–5, 101, 220
Eata 151, 153–4, 214–15
Ebrauc 219
Ecgfrith, King 153, 241–5

Ecgred, Bishop of
    Lindisfarne 265–6
Eden, River 124–6
Ednam 125–6, 128–9
Edward I, King 132
Edward II, King 134
Edwin, King 180, 219
Eildon Hills 23, 25–8
Elliot, Walter 66–7, 68, 69
Emmanuel Head 225–6
Erskine, David Stuart, Earl
    of Buchan 41, 52
Etal 151, 154, 155
Ethelwald 250–1
Ettrick Forest 23

farming 26–8, 127–8
Farne Islands 9; *see also*
    Inner Farne
Fenwick 168–9
Ferguson, Johnny 237
Flambard, Ranulf 270
Flodden 135, 141, 156–9
Ford 155–7
Forfarshire (steamer) 286–7,
    288
France 60, 156
Francis of Assisi, St 80
Friga 101

Gaullauc, King of Elmet
    171, 172–3
Gentiles, the 99, 179

ghost fence 26–7
Gododdin 99
Gospels 252; see also
    Lindisfarne Gospels
Grandstand 161–2
Great Dirrington Law 23
Great Heathen Army 195,
    233, 266, 267
Gregory the Great, Pope
    48
Guthfrith, King 268

Haig, Alexander 41
Haig, Douglas, Earl 51
Halfdan 266, 267, 268
Haliwerfolc (Holy Man's
    People) 278
Hallowe'en 26–7
heads 26–7
Hebrides 263
henges 67–8, 69
Henry VIII, King xvi, 156,
    259
Herefrith 293–4, 297, 301,
    302
Heugh, the 197–201, 205–6
Hexham 242–3
Higbald, Bishop of
    Lindisfarne 262–3
Hinnegan, Willie 236–7
Hobthrush (St Cuthbert's
    Isle) 213–14, 216–17, 220–2,
    233–4

Holburn 165–6
Holloway, Richard 89
holy ground 64–5
Hope-Taylor, Brian 100
horses 81–4, 248–9
Horton Moor 162–3
Hroc 118
Hudson, Edward 259, 260
human sacrifice 49, 100
Hussa 172–3
Hywel Dda 176

Ida 173, 218
Imbolc 27, 48
immersion 85–6
Inner Farne 12, 79–80,
    240–1, 283, 289–97
inscriptions 201–4
insular majuscule script
    254
Iona xv, 263
Ireland 91–2, 264
Islandshire 154–5, 278
Ivanhoe (Scott) 50

James IV, King 141, 156,
    157–8
Jekyll, Gertrude 259–60
Jesus Christ 122
John, St 113, 114, 252
John Baliol, King 132
Johnston, Dr George 15
Justinian, St 44

Kellaw, Richard 134–5
Kelso 118–22

Lammermuir Hills 23
Latinus 180
Law 111–15
Lewis 223–4
Life of St Cuthbert 16–18, 29,
    152
Lindisfarne xii, xiv–xvii,
    4–13
    and Castle 257–61
    and desertion 270–1
    and shires 154–5
    and Vikings 261–3, 265–6
    see also causeway; tide
Lindisfarne Gospels xv–xvi,
    194–5, 196–7, 249–56, 267
    and Chester-le-Street 268
    and Vikings 262
Links 234–5, 237–40
Little Beblowe 231
Little Dirrington Law 23
Llofan Llaf Difo 175
Longstone lighthouse 9–10,
    286
Lughnasa 27, 48
Luke, St 252
Lutyens, Sir Edwin 259, 260

McCombie, Jim 210
MacIver, Donald 223,
    224–5

Madron Well 106
Maiden Paps 23
Makerstoun House 111–13
Malcolm II, King 130–1
Malcolm III Canmore, King
    60
Margaret, Queen 132
Mark, St 252
Marmion (Scott) 144
Martin of Tours, St 180
Mary, St 150–1
Marygate (Lindisfarne)
    231–2
Matthew, St 252
Maxton Kirk 109–10
Meeting Fort 23
metal deposits 105–6
miracles 152–3
missionaries 48–9
Modan, St 47–8, 49, 50
Modranect (Mothers'
    Night) 101
Moffat family 69–75, 96–7,
    126–9, 272–5
monasteries 41, 60, 64–5,
    180–1, 259
Monks' Road 52–3
Monksford 52, 53–5
Morcant Bwlch 171, 172–3
Mortimer, Sir Alan 65

name stones 232
Nennius (Ninya):

Historia Brittonum 172–3,
   175, 180
Newtown St Boswells 64
Ninian, St 102
Norham 265–6
Norhamshire 155, 278
Northumberland 99
Norway 202–3, 265

Old Melrose xi, xii, xv, 12,
   32–7, 57–9
  and Cuthbert 62–4, 86–8
  and Drythelm 85–6
  and history 60–2, 65–9
orienteering 20–1
Osborne's Fort 259
Osred, King 152
Oswald, King xv, 219
Oswiu of Northumbria,
   King 94–5
Oswy, King xv, 265
Ouse bay 257, 261
Owain, King 130–1, 176

paganism 26–7, 29, 48–50,
   99–100, 101
Pascal, Blaise 228–9
Passover 94
Picts 102, 117, 245
Pilgrim Poles 182
pilgrims 256–7
Pilgrims Coffee House 187,
   222, 281

Plague of Justinian 76, 87
politics 241–5
Pontin, Tommy 210
Popple Well 234
Prayer Holes 199–200
priory 277, 278–81
propitiation 49, 106
psalmody 238–9

Reddenburn 130
Rheged 171, 172–4, 180
Rhun 180
Richard de Morville of
   Lauder 65, 68
Riderch Hen, King of
   Strathclyde 171, 172–3
rituals 105–6
Robert Bruce, King 134
Roman Empire xiv, 26
Roxburgh Castle 116–18
Ruberslaw 23
Ruthwell Cross 191–2

St Andrews University 96
St Baldred's Boat 144
St Boisil's Chapel 107, 108
St Boswells 64, 107–8
St Bridget's Day 27
St Cuthbert's Cave 167–8
St Cuthbert's Chapel 143–4
St Cuthbert's Way 109,
   168–9
St Mary's Chapel 150–1

St Mary's Church
  (Lindisfarne) 189–91,
  192–3, 194–7, 210–13
saints 49–50, 152–3
Samhuinn 26, 48
Saxons xiv, 49, 76, 176, 179
Scott, Walter 41, 50–1, 144
seafaring 91–3
Seahouses 285
Second Creation 76, 80, 303
Second World War 170–1,
  202–3
Severin, Tim 91–2
Ship Inn 204, 205
shires 154–5
Sicga 263
Sime, Rev. W.L. 16, 35
slavery 264–5
Sons of Death xvi
*Statistical Account for
  Scotland* 26
Stephanus Eddius 152
Stevenson, Robert Louis 273–4
stonemasonry 236–7
Surrey, Earl of 156, 157–8
Symeon of Durham 131,
  266–7
Synod of Whitby 95–6, 101,
  151

Taliesin 177–9
Theodore, Archbishop of
  Canterbury 244

Theodoric 172–3, 175
Thomas of Ercildoune 23,
  40–1
Thor Longus 125
Thunor 101
tide 4–5, 181, 188–9
Till, River 141–50
Tiw 101
tonsure 48, 95
Tweed, River 23, 24, 84–6,
  104–5
  and Kelso 119–22
  and the Law 111–15
  and Monksford 52, 53–5
  and sailing 93
Twizel Bridge 141–2

Urien Yrechwydd, King of
  Rheged 171, 172–4, 175–6,
  177–8

Venutius, King 26
Vikings xvi, 144, 261–6,
  267–8
visions 228–9

Wales 176, 179–80
Wallace, William 42
war 170–1
Wark Castle 133–6
Wars of Independence 118,
  132
water divining 66–7

Waterford, Lady 155–6
Weir, Doddie 305–6
White Church of Stone 269
Whitmuirhaugh 123–4
Wilfred, St 151–2, 154, 256
William de Carilef, Bishop 269–70
William I the Conqueror, King 269, 271, 277
Woden 101

wood henges 67–8, 69
Wrangham 16, 17, 18, 32–3

Y Gynt 99
Y Mab Darogan 176
Yeavering Bell 23
Yeholmshire 155
Yellow Plague 87–8, 90, 95
Yrfai, Lord of Edinburgh 179